SPRINGER
LAB MANUAL

W0055399

Springer-Verlag Berlin Heidelberg GmbH

Laura Caponi · Paola Migliorini (Eds.)

Antibody Usage in the Lab

With 22 Figures

Springer

L. CAPONI AND P. MIGLIORINI
University of Pisa
Department of Clinical Medicine
Clinical Immunology Unit
Via Roma 67
56126 Pisa
Italy

Original pictures drawn by the authors.

ISBN 978-3-662-03944-1

Library of Congress Cataloging-in-Publication Data
Antibody usage in the lab / [edited by] Laura Caponi, Paola Migliorini.
    p. cm. – (Springer lab manual)
    ISBN 978-3-662-03944-1   ISBN 978-3-662-03942-7 (eBook)
    DOI 10.1007/978-3-662-03942-7

    1. Immunoassay – Laboratory manuals. 2. Immunoglobulins – Laboratory manuals.
    I. Caponi, Laura, 1962- . II. Migliorini, Paola. III. Series.
    QP519.9.I42A57   1999
    616.07'56 – dc21   99-11672

© Springer-Verlag Berlin Heidelberg 1999
Originally published by Springer-Verlag Berlin Heidelberg New York in 1999

Production: PRO EDIT GmbH, Heidelberg
Typesetting: Mitterweger Werksatz GmbH, Plankstadt
Cover design: design & production GmbH, Heidelberg
SPIN: 10682977   27/3136 – 5 4 3 2 1 0 – Printed on acid-free paper

Table of Contents

List of Contributors

LAURA CAPONI
Department of Internal Medicine, Clinical Immunology Unit
University of Pisa
Ospedale S. Chiara
Via Roma 67
56126 Pisa, Italy

GIANCARLO CARBONE
Department of Experimental Medicine, Histology Unit
University of Genoa
Via de Toni 4
16132 Genova, Italy

MARINA FABBI
National Institute for Cancer Research and
Advanced Biotechnology Center
Largo R. Benzi 10
16132 Genova, Italy

DANIELA LUCCHESI
Institute of Mutagenesis and Differentiation,
CNR, Peptide Synthesis Laboratory
Via Svezia 2A
56124 Pisa, Italy

PAOLA MIGLIORINI
Department of Internal Medicine, Clinical Immunology Unit
University of Pisa
Ospedale S. Chiara
Via Roma 67
56126 Pisa, Italy

PAOLO ROVERO
Dipartimento di Scienze Farmaceutiche
Universita' di Salerno
Via Ponte Don Melillo, 11C
84084 Fisciano, Salerno
Italy

MICAELA TISO
Institute of Biological Chemistry and
Advanced Biotechnology Center
University of Genoa
Largo R. Benzi 10
16132 Genova, Italy

Chapter 1

Introduction

LAURA CAPONI

Immunoassays are techniques that use labeled or unlabeled antibodies as reagents for the detection and quantification of sample analytes. The popularity of these methods has increased because they are precise, sensitive and relatively inexpensive. In fact, antibodies have the ability to recognize their ligands with high specificity and can also be labeled fairly easily without significant loss of binding activity. Thanks to these properties, they can be used together with other reagents in analytical procedures. Furthermore, the substitution of isotopes with non-radioactive labels has made these techniques less hazardous to both users and the environment. Finally, the specificity and the high affinity or avidity of antibody binding make it an extremely sensitive technique for the detection of minute amounts of antigens.

This introductory chapter will briefly explain some of the mechanisms underlying the ability of antibodies to recognise their ligands, including the structure of antibodies, some general notions regarding the structure of antigens, and finally the basic principles governing their interactions. Only those structural aspects that may possibly influence the performance of an assay will be taken into account. These (complex) subjects will not be treated entirely and the information presented is merely intended to provide a minimum background for better understanding of the following chapters. For a more complete description of the structural and functional properties of antibodies and the mechanisms that determine their variability, the reader is encouraged to consult a standard text on immunology. This chapter will also briefly discuss some of the common factors that may influence antigen-antibody binding and, as a conse-

Laura Caponi, University of Pisa, Department of Internal Medicine, Clinical Immunology Unit, Pisa, Italy

quence, the assay itself. Specific problems relating to each technique will be discussed in the relevant chapters.

Antibodies

Antibodies are glycoproteins that constitute one of the principle tools used by the organism to defend itself against potentially damaging foreign agents. They are synthesized by and exposed on the surfaces of lymphocytes, and are then secreted into biological fluids. One of the most important features of antibodies, and the one that makes them suitable tools for assays, is their ability to recognize and bind molecules in solution.

Antibodies are quite heterogeneous molecules, but share a common basic structure that determines some of their characteristics. They are heterodimers made up of four polypeptide chains: two identical polypeptides ranging in molecular weight from 50 to 70 kDa, denominated the "heavy chains" (H); and two identical polypeptides approximately 25 kDa in molecular weight called the light chains (L). Each light chain is bound to a heavy chain and the two heavy chains are bound together. These four chains are covalently linked by disulphide bonds (see Fig. 1), while non-covalent (for example, hydrophobic) interactions play a role in determining their tertiary and quaternary structures.

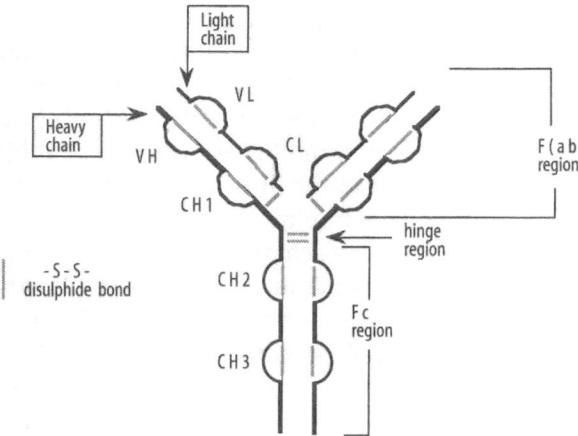

Fig. 1. The basic structure of an antibody

The amino acid sequences of the heavy and the light chains show a greater variability at their amino terminal fragments; thus, for every chain it is possible to recognise a variable portion (VH for the heavy chain and VL for the light chain) and a constant portion (CH for the heavy chain and CL for the light chain). Within each chain, the amino acid composition allows the formation of globular domains during folding. These domains are responsible for some of the biological properties of the antibody. The variable region of each heavy and light chain contains one domain. The constant region of the light chain contains only one domain, while within the longer constant region of the heavy chain three or more domains are possible.

Each variable region contains three areas, denominated the hypervariable regions, where the amino acid variability is particularly high. During protein folding and assemblage of the antibody molecule, the variable region of each light chain faces to the variable region of one of the heavy chains, and their hypervariable regions approach one another, thus creating the combinatory site for the antigen. In this way each molecule of antibody exposes two possible sites for combination with antigens of the same specificity – that is, with the ability to bind the same epitope. In other words, the basic antibody (for example, an IgG) is divalent and its specificity is based on the ability of the combinatory site to recognize a complementary region of the antigen.

The antibodies may be subdivided into five classes of isotypes, since five types of heavy chains exist in human antibodies: IgG, IgM, IgA, IgE, and IgD. In humans, the IgG and IgA isotypes can be further divided into four and two subclasses, respectively. These subdivisions are based on the sequences of the constant regions of their heavy chains, and antibodies of different isotypes possess different properties which determine their functions.

The carboxyterminal region of the two heavy chains, the Fc region, plays a principle role in certain antibody functions in vivo such as complement fixation, mast cell degranulation or placental transfer. This carboxyterminal region can be selectively separated from the antibody molecule: by proteolytic digestion, the enzyme papaine cleaves the hinge region, separating the Fc fragment from two other fragments, the F(ab) fragments, each containing an aminoterminal portion of the antibody. Each F(ab) bears a combinatory site for the antigen. The enzyme pepsin can further degrade the Fc region by attacking several sites on this portion of the antibody molecule, generating a large fragment called F(ab)₂ which contains both of the combinatory sites for the antigen. The F(ab)₂ fragment is divalent and therefore possesses the same affinity and avidity for the antigen as the intact antibody, but it lacks the effector functions of Fc. For certain assays it is preferable to use the F(ab)₂ fragment rather than the whole antibody molecule in order to avoid possible interference due to the binding of complement proteins or rheumatoid factors to the Fc fragment of the antibody.

Figure 1 shows the basic structure of an IgG molecule. Other isotypes may look slightly different: IgE and IgM, for example, each have a supplementary domain located in the carboxyterminal region. Secreted IgM is configured as a pentameric complex; a polypeptide chain called the junction (J) peptide links five molecules of antibody at the carboxyterminal region by means of disulphide bonds. In the case of IgA, the same J chain is responsible for the binding of two IgA molecules to form the dimer which is present in serum and, with a supplementary secretory chain, in secretions. When the antibody is the analyte of interest, each class can be selectively detected since each isotype exhibits different antigenic properties.

After a primary immunization, several lymphocytes carrying antibody molecules with different specificities may react with the antigen and thereby be recruited. During the course of the immune response, following subsequent exposures to the antigen, antibodies may undergo structural modifications of their amino acid sequences. This leads to modifications in the properties of the individual antibody and to an increase in the affinity of the immunologic response, due to mutations occurring in genes coding for the hypervariable regions and genetic rearrangements that cause isotype switching. The end result of these processes is an increase in the average affinity of the antibodies produced during subsequent exposures to the antigen (see below).

Antigens

An antigen can be defined as a substance able to bind an antibody or a T cell receptor. Immunogens are substances able to induce an immune response, either humoral or cellular. All immunogens may be considered antigens, but not every antigen may be an immunogen since not every antigenic substance is able to induce an immune response. An antigen that is not immunogenic may be able to elicit an immune response, however, if linked to a larger molecule (usually a protein) called a "carrier".

The portion of the antigen that is recognized by the antibody, i.e. the fragment coming into close contact with the combinatory site of the antibody, is called the antigenic determinant or epitope. Many of the substances used in assays or specific procedures may influence the recognition of the antigenic determinants.

The protein epitopes recognized by antibodies can be either linear or conformational. An epitope is linear when it is made up of amino acids located sequentially in the polypeptide chain; this means that it maintains its antigenicity even when the protein has been denatured and has lost its three-dimensional structure. Linear epitopes can be detected by means of techniques based on the use of denaturing substances such as strongly charged detergents.

On the other hand, the folding of the tertiary structure of a protein allows the formation of a conformational epitope, in which amino acids located in distant portions of the polypeptide chain can come into close contact with one another, producing antigenic determinants. Thus, a conformational epitope retains its antigenic reactivity only when the protein maintains its native conformation, and harsh treatments causing the denaturation of the protein will lead to the disappearance of the conformational epitope and consequently to the loss of its antigenicity. Sometimes when the factors that originally caused the denaturation are eliminated, a partial renaturation of the protein can take place and the antigenic determinant may again become available for binding.

Both linear and conformational epitopes may lose their antigenicity for other reasons. For example, the protein epitopes may be glycosylated or phosphorylated, and treatments that provoke the loss of glycides or posphate groups could lead to the loss of antigenicity.

Antigen-antibody binding

The binding of an antigen to an antibody is not covalent, but is rather based on electrostatic or van der Waals interactions. If close contact between the epitope and the combinatory site is possible, also hydrophobic interactions may contribute to the binding.

Antigen-antibody binding is reversible and may be influenced by different conditions such as pH and ionic strength. The pH and composition of the buffers used in various assays therefore represent important variables that must be taken into account when optimal binding is necessary, while modifications in the buffer composition can be used to disrupt these bonds. Lowering the pH or raising the salt concentration are common laboratory practices used to separate antigens and antibodies. In addition to the ionic strength and pH of the buffer, other parameters may be manipulated to affect the antigen-antibody reaction; these will be examined separately in the different chapters in the context of the specific techniques in which they may play a role.

Since their binding is reversible, a dynamic equilibrium exists between the antigen and its antibody, an equilibrium that depends on their "affinity", i.e. on the strength of the interactions between them. If the antigen contains a single epitope, it can bind to only one combinatory site on the antibody. In this case, the affinity is simply expressed by the constant of dissociation of the complex as it occurs in dynamic equilibrium reactions. Actually, besides the affinity expressed by the constant of dissociation, other elements may influence the strength of the antigen-antibody bond. One of these factors is the number of antigenic determinants present on the antigen. Another is the fact that the same antibody may establish bridge linking, since a molecule of antibody always has two identical combinatory sites and thus can bind two molecules of antigen.

Avidity may be considered as the outcome of all of these factors (the valence of the antibody, the number and location of the antigenic determinations on the antigen), and provides a measure of the global "strength" of the antigen-antibody bond. The isotype involved is also very important in determining the avidity of the antigen-antibody reaction. Binding involving a pentameric IgM molecule, which is considered pentavalent rather than decavalent due to problems of steric hindrance, will have a higher avidity than binding involving a single divalent IgG.

Large antigens will contain several epitopes that may be different or identical. In such cases the antigen is multivalent and many antibodies can bind to it. However, this multiple binding may be limited by steric hindrance, since antibodies are also large molecules and cannot bind epitopes that lie too close to one another.

By adding increasing amounts of antibody to a solution containing a fixed amount of antigen, the antibody will bind increasing amounts of the antigen and generate soluble immunocomplexes. When the optimal concentration (the equivalence point) is reached, the interaction between antigens and antibodies gives rise to a structure in which the maximum number of molecules of antigen and antibody are bound in a mesh structure called lattice, which may sometimes precipitate. Interactions among the Fc fragments of the antibodies also contribute to the formation of the lattice. The precipitating immunocomplexes can then be re-solubilized, providing an additional amount of antibody (Fig. 2).

After antigenic stimulation a polyclonal antibody response occurs, i.e. many different lymphocytic clones are stimulated by their interaction with various epitopes, and antibodies

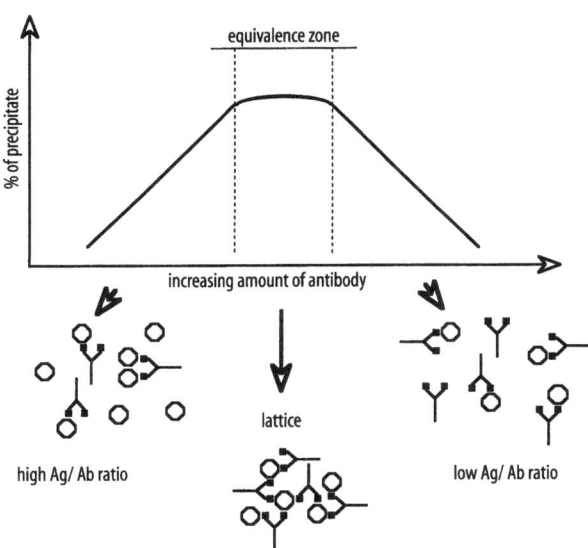

Fig. 2. Immune complexes at various antigen-antibody ratios

with different binding capacities are synthesized and secreted. If there is a second antigenic stimulation, or if the antigen exposure persists for a long time, then the genes encoding for the immunoglobulins will undergo rearrangements that lead to class switching, and mutations occur in DNA encoding for combinatory sites, leading to the production of higher affinity antibodies.

An immune serum may contain different antibody molecules directed to the target antigen, and while the affinity of each antibody may be low, the binding capacity of the whole serum represents the outcome of the individual binding capacities of the various antibodies produced by different clones. In contrast, a monoclonal antibody has a binding capacity that reflects the affinity of a single binding site directed to a single epitope. This explains why samples containing monoclonal antibodies may show a lower avidity than samples of polyclonal serum.

Chapter 2

ELISA assays

LAURA CAPONI AND PAOLA MIGLIORINI

Introduction

ELISA (Enzyme-Linked ImmunoSorbent Assay) is a well-known
and widely used laboratory technique that is able to measure the
ligands present in small amounts in biological samples. ELISA
combines a high degree of sensitivity and specificity with a detection system based on a colorimetric reaction; it can also provide quantitative results.

We will consider the most common type of ELISA, the heterogeneous non-competitive indirect ELISA. The protocols will describe in detail a common situation in which the analyte to be
detected is an antibody directed to the coated antigen (Fig. 1). In
heterogeneous assays, the reaction takes place in a liquid environment with one of the reagents immobilized on a solid phase. This
allows the removal of unbound reagents in a series of washing steps.
The term "indirect" refers to the fact that the analyte is revealed not
directly but rather by means of an intermediate reagent.

The procedure described below illustrates most of the principles involved in the setting up of an ELISA assay; the antigen is
coated onto a solid phase, samples containing the antibodies directed against the antigen are added, and the detection is carried
out by an enzyme-labelled anti-antibody. The enzyme acts on an
appropriate substrate, releasing a coloured compound that can
be easily detected by a spectrophotometer.

The information that follows can be easily adapted to cases
where the analyte to be detected is not an antibody (Fig. 2).

Laura Caponi, University of Pisa, Department of Internal Medicine, Clinical
Immunology Unit, Pisa, Italy
Paola Migliorini, University of Pisa, Department of Internal Medicine,
Clinical Immunology Unit, Pisa, Italy

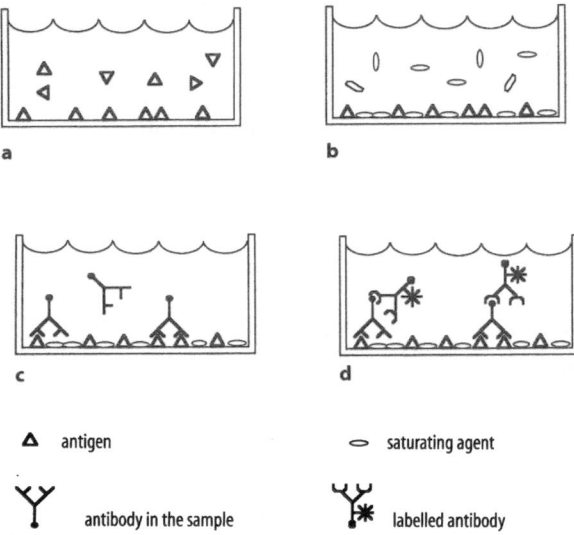

Fig. 1. Indirect ELISA:
The analyte to be detected is an antibody directed to the coated antigen.
a: coating of the antigen; b: saturation; c: adding the sample; d: adding the
labeled antibody.

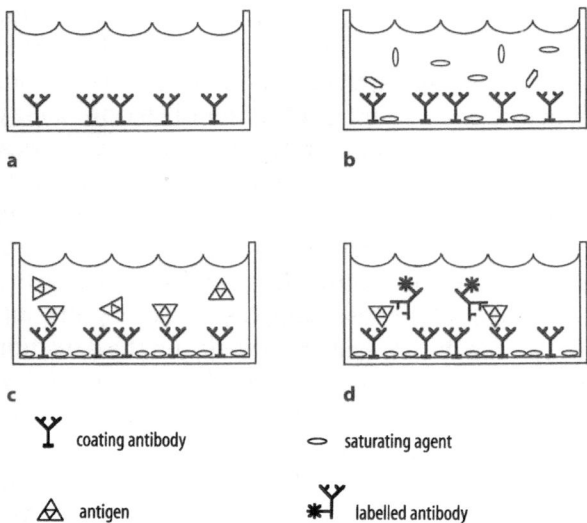

Fig. 2. Indirect (sandwich) ELISA:
The analyte to be detected is an antigen with multiple epitopes.
a: coating of the antibody; b: saturation; c: adding the sample; d: adding the
labeled antibody.

Solid phase support and coating of the antigen

The most convenient solid support for the ELISA reaction is the 96-well microtiter plate. It comes in a standardised form designed for use with most microtiter plate readers. Usually it is made of transparent polystyrene, which allows the transmission of light with low background. The relatively low binding capacity of this material can be enhanced by physical or chemical treatment, although this will often also increase the background.

In some cases the coating of the antigen to the polystyrene is indirect, i.e. it is mediated by other substances in a pre-coating step that can be carried out in the laboratory to ensure a good coating. In the case of DNA the binding to the plate can be mediated by polylysine, but for other types of antigens enhancement is obtained by pre-coating the wells with avidin and then adding the biotinylated antigen. If the coating is made with antibodies of a given specificity, then pre-coating with protein A – which can position the antibodies in such a way as to expose their binding sites – may be useful. However, a satisfactory result is often obtained by direct coating, that is by passive absorption onto the plate. In this case the forces involved in the binding are mainly electrostatic, although hydrophobic bonds may also be involved. Covalent bonds are formed only when chemically modified plates are used.

For direct coating, proteins and peptides are diluted in a suitable medium; three buffers with different pH values (acid, neutral and basic) should be tested in order to determine the best coating condition. In our experience, most proteins coat efficiently at basic pH in a carbonate/bicarbonate buffer and most peptides coat well in PBS at physiological pH. Other antigens with different chemical structures may require a non-aqueous medium to achieve optimal coating; for example, lipid antigens are usually suspended in chloroform or ethanol and the solvent is allowed to evaporate from the plates.

The time required for coating depends on the temperature: 1 hour at 37°C is often sufficient, but it is also common practice to add the coating solution and leave the plates overnight at 4°C. The volume of antigen solution used is usually 50 to 100 µl/well. The choice of coating volume will determine the subsequent volume of the sample. The amount of antigen for direct coating usually ranges from 1 to 10 µg/ml for peptides and from 0.5 to 5 µg/ml for pure proteins.

The best procedure to establish the optimal amount of antigen for the coating is to test increasing concentrations of antigen in the coating buffer. The amount of antigen absorbed on the solid phase can then be measured either directly, if the antigen itself is labeled, or indirectly using a specific antibody. The amount of antigen adsorbed to the solid phase increases proportionally with the antigen concentration in the coating solution, and then levels off (Fig. 3). The first antigen concentration at which this plateau is reached is usually optimal for direct binding. It is not a good idea to choose a higher concentration since in such cases the antigen molecules (especially if the antigen is a protein) may interact among themselves, producing a multi-layer that could undergo desorption during subsequent steps. For inhibition assays in which two ligands compete for binding to the antigen on the solid phase, a lower antigen concentration (from 50 to 75% of the plateau level) is usually better.

Saturation

After the absorption of the antigen onto the plate, the coating solution is discarded and a saturating agent is added to the wells in order to saturate those areas of the plate not covered by the antigen. In this way any non-specific binding of the antibody in the sample to the plates (i.e. by means of the same forces involved in the coating) will in principle be prevented. In any case the

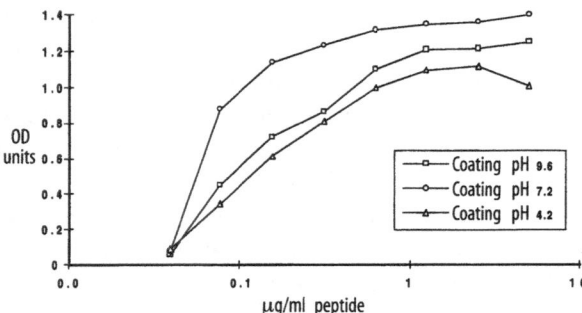

Fig. 3. Coating a peptide at different pHs.
The best coating is obtained at pH 7.2. The plateau is reached around the concentration of 1 μg/ml.

binding of the sample and the labeled antibody to the saturating agent must always be evaluated by testing the binding of the samples to uncoated but saturated plates.

A number of saturating agents are available. In our experience 3% bovine serum albumin (BSA) is reliable for most assays, but sometimes other less expensive agents can be used. For example, in certain situations 5% non-fat milk or 2% casein may reduce the background even more than BSA. Another good saturating agent is 1% porcine gelatine. Casein and gelatine both have good saturation capacities because they are composed of a mixture of proteins of different sizes that can efficiently cover those areas left uncovered by the antigen. Our laboratory has verified that the saturating ability of gelatine significantly increases after autoclaving, a step that probably splits the larger molecules into smaller fragments. A 5% solution of gelatine can be prepared, autoclaved, and then stored, with the addition of 0.02% sodium azide, at room temperature as a stock solution.

Whichever saturation agent is chosen, it must be diluted with the same buffer as that used for the sample, but no detergent is added at this step since it may interfere with the binding of the saturation agent to the plate. One exception is Tween 20, a mild non-ionic detergent that also seems to have saturation properties. The volume of the saturation solution must always be greater than the volume used for the coating. We usually add 100 µl/well for a 50 µl/well coating and 200 µl/well for a 100 ul/well coating. The saturation step is generally carried out at room temperature for 1 hour.

Sample dilution

Once the concentration of the antigen for the coating is chosen, the dilution of the sample must be optimized. Several dilutions of positive and negative samples should be run in order to obtain binding curves, and the optimal dilution should be chosen from the linear portion of the binding curve, since in this range the assay has the highest sensitivity – that is, the highest colorimetric difference in relation to the differences in the amount of ligand. At the chosen dilution, the negative samples should still bind poorly to the coated plates. If this is not the case, perhaps a less than optimal saturating agent has been chosen.

The diluting buffer is usually a physiological solution such as PBS, although Tris-HCl buffer at pH 7.4 can also be used, especially if the label is alkaline phosphatase, since this enzyme is inhibited by phosphate groups. Usually a low concentration of the saturation agent is used in the diluting solution (e.g., 1% BSA or 0.5% gelatine), and a detergent is added (Tween 20 0.05%, or some other mild detergent such as Triton X-100 or CHAPS) to prevent non-specific binding. The volume of the sample must be the same as that used for the coating and the samples should be tested in duplicate.

Labeled antibody and substrates

The detection system consists of an antibody labeled with an enzyme and directed to the analyte of the sample, plus its substrate. The substrate, when enzymatically cleaved, releases a coloured compound that can be detected by a spectrophotometer.

Generally speaking, the product of the enzymatic reaction must be proportional to the analyte in the sample, and the speed of the reaction should be linear over time. Under the conditions described above, the maximum speed of the enzymatic reaction (i.e. the amount of substrate transformed per time unit) is obtained when the enzyme is saturated by the substrate. All of these requirements are usually satisfied under optimal conditions.

Since a single enzyme molecule can act on several molecules of substrate, the catalyzed reaction is amplified and this enhances the sensitivity of the assay. However, it must kept in mind that the enzymatic reaction may be influenced by temperature (as, for example, in the "edge effect" described below) or by interference from substances contained in the sample.

The most frequently used enzymes with their corresponding substrates are: alkaline phosphatase with para-nitrophenylphosphate (PNPP), horseradish peroxidase with o-phenylenediamine dihydrochloride (OPD), and (less frequently) β-galactosidase with orto-nitrophenyl-β-D-galactopyranoside (ONPG). None of these enzyme labels seem to offer significant advantages over the others and therefore one's choice may be based on laboratory tradition and personal preference. A number of labeled antibodies are commercially available; the manufacturer's instructions provide reliable guidelines to their use, although sometimes the working dilution may have to be adjusted.

The labeled antibody, e.g. an anti-human IgG antibody, is diluted in the same dilution buffer as the sample, and a volume identical to that of the sample volume is dispensed into the wells.

A fresh batch of substrate solution must be prepared for each assay. In some cases, a blocking solution may be added to stop the enzymatic reaction.

Washing and incubation

The samples and the second-step antibody are usually incubated for 2 to 4 hours at room temperature by leaving the plate on a gentle orbital shaker. Alternatively, the plate can be incubated overnight at 4°C: in this case, the longer incubation time compensates for the slower kinetics of the reaction at low temperature.

Before the addition of the detection antibody and again before the addition of the substrate, a thorough washing step is essential. The washing solution often contains a mild detergent (usually Tween 20) to lower the background (see protocols). Generally no washing is required after the coating step, but if washing steps are required after the coating or pre-coating steps, no detergent should be used in the solution since this could interfere with the subsequent saturation step.

The volume of the washing solution should be the same as that of the saturation step or even higher. Unless otherwise specified, no incubation of the washing solution in the plates is required. If necessary, a higher number of washings, a higher concentration of detergent in the washing buffer, or the incubation of the washing solution in the wells for 5 minutes can help to reduce the background.

Expressing results

The ELISA technique allows to distinguish negative from positive samples and to measure the ligand concentration in the samples. In order to establish the threshold separating normal from pathological specimens, a number of samples (at least 50) drawn from healthy subjects (for example blood donors) should also be tested and the results statistically evaluated. In most cases, the results from a healthy population display a Gaussian distribution and the cut-off can be established as the mean plus 2 standard deviations (or 3 SD to include 98% of the possible normal results). If the ligand of the ELISA is present in the normal samples, this evaluation will establish the upper limit of normal values. If the ligand of the ELISA is normally not present in the sample (e.g. an autoantibody), the upper limit of normality indicates the normal human sera background of the assay.

If a "yes/no" result is not sufficient and quantitation is required, a reference curve can be drawn up based on the same positive sample (if available). At least three dilutions of the positive sample must be included for each test. The dilutions should be chosen from the linear segment of the binding curve. Usually the units of analyte can be easily expressed as the reciprocal of the sample dilution (on the x-axis). For a large range, semilog plotting is preferable and a reference curve for the unknown samples is thereby obtained (Fig. 4). Another way to express the results is in terms of the percentage of binding with reference to a high positive sample.

Even if quantitation is not required, at least one positive sample (but preferably three with different binding capacities – high, medium and low) should be included in the test as a quality control, as well as at least one negative sample and a control well containing all of the reagents but no sample, in order to verify the performance of the test and the background levels ascribable to aspecific binding of the labelled antibody.

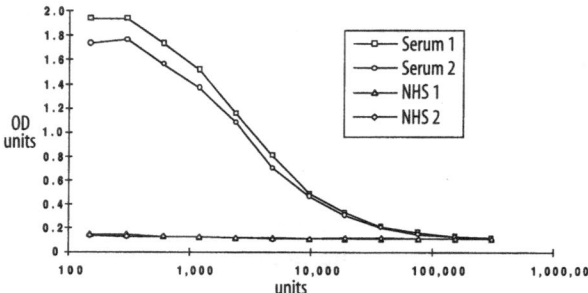

Fig. 4. Binding curve for ribosomal peptide coated on the plate.
Serum 1 and serum 2 contain anti-ribosomal antibodies. NHS 1 and NHS 2 are sera from healthy blood donors.

Common amplification systems

Different variations on the basic scheme described above are possible; while the enzymatic system usually produces a good signal, modifications can improve the detection limits of the technique.

The sensitivity of the ELISA may be enhanced by using the avidin-biotin system. Avidin is a basic glycoprotein obtained from egg white that has a high affinity for biotin, a small vitamin that conjugates easily to proteins, usually without altering their properties. In most cases streptavidin (a protein similar to avidin but produced by bacteria) is preferable to avidin because it has a more favourable pI (5.5 – 6.5) and is not glycosylated, thus avoiding possible aspecific reactions. The affinity of the (strept)avidin-biotin system is much higher than that of the antigen-antibody complex. Moreover, four molecules of biotin can be bound by a single molecule of (strept)avidin. These two factors can help to raise the sensitivity of the assay without a significant increase in background. Usually antibodies of a given specificity can be biotinylated in a simple procedure without loss of binding capacity. Then (strept)avidin linked to the enzyme label is added, and finally a chromogenic reaction is carried out. Variations on this procedure are also possible.

Chemiluminescence is another detection system that offers certain advantages. It may have a higher sensitivity than conventional ELISA since it does not rely on substrates producing coloured compounds, but rather on a chemical reaction producing an unstable, excited intermediate state which emits light at a specific wavelength as it returns to the ground state. One of the most reliable of these systems is based on peroxidase/H_2O_2, which is able to oxidize luminol and its derivatives, but other substrates are available and other enzymes may be used. Chemiluminescence has became more widely used ever since the introduction of various improvements, such as the addition to the substrate medium of substances designed to enhance and stabilize the emitted signal. These serve to speed up the reaction time, and to create a stronger signal that is stable for a reasonable length of time, thus allowing repeat readings if necessary. Chemiluminescence requires specific equipment, including solid supports to reflect the emitted light and a luminometer (a luminescence reader) to measure the signal.

Drawbacks of the ELISA

Various factors may negatively affect the results of the ELISA test even under standardised conditions. Here we will review some of the more common problems encountered.

The adhesion of the antigen to the solid support can vary; therefore one should always control the coating conditions (time and temperature) and the composition and freshness of the buffers. It must also be kept in mind that even if one always uses exactly the same type of microtiter plate, its coating efficiency may vary between batches.

If your control well, i.e. the saturated, antigen-coated well containing buffer alone, exhibits a developed substrate, there are various possible explanations. If the uncoated wells also produce coloured compounds, then the saturation of the uncoated areas of the plastic may have been incomplete and a better saturation compound for that type of plate must be found. If the uncoated but saturated plate does not develop a significant OD, an interfering reaction between the second-step antibody and the saturation agent or the antigen (e.g. an electrostatic interaction) may be the cause. Sometimes the use of $F(ab)_2$-labelled antibodies rather than complete antibodies can decrease this type of aspecific binding.

If the development of the substrate is extremely low or even non-existent in all the wells, including your positive control, it is possible that the enzyme linked to the second-step antibody has lost its activity. In such cases this problem may appear quite abruptly. The quickest way to test this possibility is to borrow another second-step antibody from a colleague and to re-run the assay on a small scale. One particular drawback is the "edge effect", that is, the development of the substrate may occur more slowly (or, on occasion, more quickly) in the peripheral wells than in the central ones. Usually this happens because the peripheral wells have been exposed to a different temperature than the central ones and this may significantly influence the rate of enzymatic activity. For example, when you are using a warm substrate, the central wells will be protected by the surrounding ones, while the heat will be dispersed more rapidly from the peripheral wells, thus slowing the enzymatic reaction. This effect may become even more marked when many plates are stacked one on top of the other. On the contrary, if cold substrate is added and the plate is then incubated at 37°C, the peripheral wells will warm more quickly than the central ones.

Subprotocol 1
ELISA for anti-ribosomal antibodies

■ ■ Materials

Equipment

- Flat-bottomed, 96-well microtiter plates. We use Maxi-Sorp F96 plates (NUNC, Denmark) for proteins and peptides, but other brands may be equally suitable.
- Pipettemen and tips.
- Multi-channel pipette with adjustable volumes (50-200 µl).
- Magnetic stirrer.
- Rocking shaker.
- Spectrophotometric plate reader equipped with a 405-nm wavelength filter.

Reagents

- Salts: Na_2HPO_4, KH_2PO_4, NaCl, KCl, $MgCl_2$, Na_2CO_3, $NaHCO_3$, NaN_3
- Gelatine (e.g. Merck)
- Tween 20
- Alkaline phosphatase-labeled detection antibody: goat anti-human IgG (e.g. Sigma).
- para-NitroPhenylPhosphate (PNPP) tablets or powder

Solutions

- PBS buffer
- 5% gelatine
- 10% NaN_3
- PBS - 1% gelatine for the saturation step

- PBS - 0.05% Tween 20 - 0.5% gelatine for the dilutions
- PBS - 0.05% Tween 20 for the washing step
- Carbonate/bicarbonate buffer pH 9.6 for the substrate
- MgCl$_2$ 1M for the substrate
- Substrate solution

Preparation

- **PBS buffer**
  Dissolve 8 g of NaCl, 0.2 g KCl, 1.44 g Na$_2$HPO$_4$, 0.24 g KH$_2$PO$_4$, adjust the volume to 1 l with distilled water. Store at 4°C.
- **NaN$_3$ 10%**
  Dissolve 0.1 g of NaN$_3$ in 1 ml of water. Store it at room temperature. **Warning: NaN$_3$ is a poison.**
- **Gelatine stock** must be prepared in advance:
  Dissolve 5 g of gelatine in 100 ml of PBS in a pyrex beaker. Dissolve on a hotplate using a magnetic stirrer. Pour the solution into a pyrex bottle and autoclave at 121°C for 20 minutes. Add 2 µl of NaN$_3$ 10% for each ml of gelatine (0.02%). This stock solution can be stored at room temperature. The PBS - 1% gelatine saturation solution may be prepared just before use.
- **PBS - 1% gelatine for the saturation step**
  The amount needed for 1 microtiter plate is 15 ml: add 3 ml of gelatine stock to 12 ml of PBS buffer.
- **PBS - 0.05% Tween 20** for the washing step
  Add 50 µl of Tween 20 to 100 ml of PBS. Store at 4°C.
- **PBS - 0.05% Tween 20 - 0.5% gelatine** for the dilutions
  Add 1 ml of gelatine stock to 9 ml of PBS - 0.05% Tween 20.
- **Carbonate/bicarbonate buffer pH 9.6 for the substrate**
  Dissolve 1.59 g of Na$_2$CO$_3$ and 2.93 g of NaHCO$_3$ in distilled water. Adjust the volume to 1 l. Store at 4°C.
- **MgCl$_2$ 1M**
  Add 0.952 g of anhydrous MgCl$_2$ to 10 ml of distilled water.
- **Substrate solution:** the amount needed has to be prepared freshly:
  Add 1 mg (one tablet corresponds to 5 mg) of PNPP for each ml of carbonate buffer. Add 2 µl of MgCl$_2$ 1M for each ml of carbonate buffer. Mix well on a vortex. If the plate is going to be developed at 37°C, heat the prepared substrate solution before incubating the plate.

Procedure

1. Dilute the peptide from the stock solution to the appropriate concentration for the coating (we use 1 µg/ml). Deliver 50 µl per well. Cover with parafilm to prevent evaporation and incubate at 4°C overnight.

Coating of the plate

Note: Peptide stock solution (e.g. 2 mg/ml) is aliquoted and stored at -20°C.

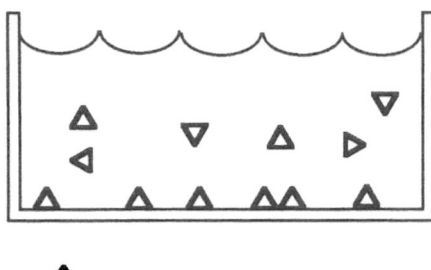

△ antigen

2. Discard the coating solution. Wash once with PBS buffer (50 µl per well), empty the plate and tap it on a paper towel. Add the PBS - gelatine 1% solution to both the coated wells and to the uncoated control wells (150 µl per well). Incubate for 1 h on a rocking platform at room temperature.

Saturation of the plate

Note: The saturation solution should be prepared fresh from an autoclaved 5% stock gelatine preparation, with NaN_3 added as preservative.

Note: The stock solution can be stored at room temperature.

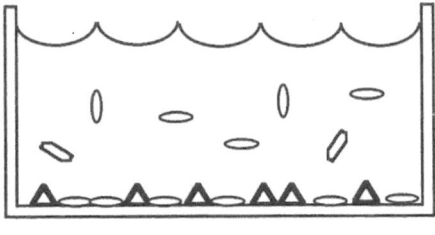

⬯ gelatine

Adding the samples

3. Discard the saturation solution. Add the samples (i.e., human sera diluted 1:300 in PBS - gelatine 0.5% - Tween 20 0.05%) in duplicate to the coated and uncoated wells (50 µl per well). Include a positive and a negative control, plus a blank (i.e., a well containing diluting buffer but no sample). Incubate for 3 hours on a rocking platform at room temperature.

Note: The appropriate sample dilution must be determined beforehand.

Note: It may be helpful, especially when processing a large number of samples, to record the distribution of the samples.

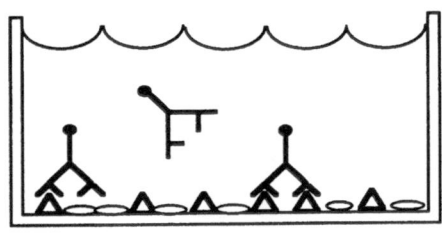

antibody in the sample

1st washing step

4. Discard the samples. Wash twice with PBS-Tween 20 0.05% (150 µl per well). Wash once with PBS (150 µl per well). After every washing empty the plate and tap it on a paper towel.

Addition of the labelled antibody

5. Add the detection antibody (in this case anti-human IgG labelled with alkaline phosphatase) at the appropriate dilution in PBS-gelatine 0.5% -Tween 20 0.05% (50 µl per well). Incubate for 3 hours at room temperature on a rocking platform, or overnight at 4°C.

Note: The appropriate dilution of the labeled detection antibody should be prepared beforehand. The manufacturer's recommendations may serve as general guidelines, but the optimal dilution must be optimized for each batch.

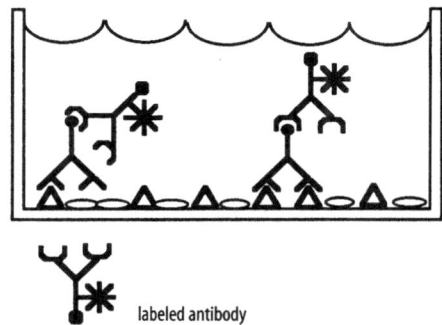

labeled antibody

6. Discard the labeled antibody. Wash twice with PBS-Tween 20 **2nd washing**
   0.05% (150 µl per well). Wash once with PBS (150 µl per well). **step**
   After every washing, empty the plate and tap it on a paper
   towel.

7. Add 80 µl substrate per well and allow the colour to develop, **Chromogenic**
   covering the plate with aluminium foil or keeping it in the **reaction**
   dark since the reaction products are photosensitive. The plate
   should be read in a spectrophotometer at 405 nm, subtracting
   the blank value.
   To obtain the final values, subtract the ODs of the samples on
   the uncoated plates from the corresponding ODs of the samples on the coated wells.

Note: The substrate solution must be prepared fresh just before
use. If necessary, the solution can be warmed before it is added to
the plates in order to accelerate the enzymatic reaction and thus
the development of colour.

Subprotocol 2
ELISA for anti-DNA antibodies

Recommendations

This procedure is suitable for the detection of anti-dsDNA or anti-ssDNA antibodies. ssDNA can be obtained by boiling dsDNA for 5 min and then cooling immediately on ice. Since an unpredictable amount of dsDNA may undergo denaturation on the plate (i.e. single-stranded tracts may be exposed), the ELISA procedure employing dsDNA may also measure some anti-ssDNA antibodies.

▓▓ Materials

Equipment

- Flat-bottomed, 96-well microtiter plates. We use Greiner, but other brands may be equally suitable.
- Multi-channel pipette with adjustable volumes (50-150 µl).
- Spectrophotometric plate reader equipped with a 405-nm wavelength filter.
- Pipettemen and tips.
- pH-meter
- Vortex and magnetic stirrers
- Rocking shaker
- DNA (for preparation see separate protocol)

Reagents

- Salts: Tris, Na_2HPO_4, KH_2PO_4, NaCl, $MgCl_2$, Na_2CO_3, $NaHCO_3$
- HCl 37%
- Tween 20
- Polylysine
- Polyglutamic acid
- Bovine serum albumin (BSA)
- Fetal calf serum (FCS)
- Labeled antibody (e.g. for human samples anti-human IgG).
- PNPP Tablets or powder

Solutions

- Phosphate-buffered saline (PBS; pH 7.3); Na_2HPO_4 50 mM; NaCl 100 mM; KH_2PO_4 10 mM
- Tris-buffered saline (TBS; pH 7.3); Tris 10 mM; NaCl 150 mM
- 5% polylysine in PBS
- Polyglutamate solution 50 µg/ml
- PBS - 3% BSA - 5% FCS for the saturation step
- PBS - 1.5% BSA - 2.5% FCS for the dilutions
- Carbonate/bicarbonate buffer pH 9.6 for the substrate
- $MgCl_2$ 1M for the substrate
- substrate solution

Preparation

- **PBS:**
  dissolve 1.42 g Na_2HPO_4, 8.76 g NaCl and 13.6 g KH_2PO_4 in 900 ml distilled water, adjust to pH 7.3 with HCl 37%, add distilled water to the final volume of 1000 ml.
- **TBS:**
  Dissolve 1.21 g Tris and 8.76 g NaCl in 800 ml distilled water, adjust to pH 7.3 with HCl 37%, and add distilled water to a final volume of 1000 ml.
- **PBS - Tween 1%:**
  Add 5 ml Tween 20 to 495 ml PBS. Dissolve using a magnetic stirrer. Store at 4°C.
- **Carbonate/bicarbonate buffer**, pH 9.6, for substrate:
  Dissolve 1.59 g Na_2CO_3 and 2.93 g $NaHCO_3$ in distilled water. Adjust the volume to 1 l. Store at 4°C.
- **PBS - BSA 3% - FCS 5%:**
  Add 3 g BSA and 5 ml FCS to 95 ml PBS. Dissolve using a magnetic stirrer. This stock solution can be prepared in large amounts, then aliquoted and stored at -20°C, thawing the amount needed just before use.
- **PBS - BSA 1.5% - FCS 2.5% - Tween 0.05%:**
  Add 1.5 g BSA, 2.5 ml FCS and 50 ul Tween 20 to 97.5 ml PBS. Dissolve using a magnetic stirrer. This stock solution can be prepared in large amounts, then aliquoted and stored at -20°C, thawing the amount needed just before use.

– **Substrate solution:**
The amount needed must be prepared fresh each time. Add 1 mg (one tablet corresponds to 5 mg) PNPP for each ml of carbonate buffer. Add 2 µl $MgCl_2$ 1M for each ml of carbonate buffer. Mix well using a vortex stirrer. If the plate is going to be developed at 37°C, heat the substrate solution before adding it to the microtiter plate.

▪ ▪ Procedure

Pre-coating of the plate

1. Dilute polylysine from the commercial preparation to 5% in PBS. Add 50 µl per well. Cover with parafilm to prevent evaporation and incubate with gentle shaking for 30-45 minutes at room temperature.

Note: This step is necessary because the binding capacity of DNA to plastic is poor. Polylysine will provide a basic surface for a better coating of the DNA.

Note: Pre-coating with poly-Lys and post-coating with poly-Glu can be substituted for by pre-coating with methylated BSA (20 µg/ml diluted from a stock solution of 1 mg/ml in H_2O).

Washing step

2. Discard the polylysine. Add 150 µl TBS per well, then empty the plate tapping it on a paper towel to remove the excess solution. Repeat this step twice.

Coating with DNA

3. Dilute the DNA to 10 µg/ml in TBS - EDTA 10 mM. Add 50 µl per well. Cover with parafilm to prevent evaporation and incubate, with gentle shaking for 2 hours at room temperature or overnight at 4°C.

Note: The DNA must be prepared in advance and stored at -20°C (see separate protocol).

Note: Add TBS-EDTA 10 mM to the non-coated control plates.

Post-coating of the plate

4. Dilute the polyglutamate from the commercial preparation to 50 µg/ml in PBS. Add 50 µl per well. Cover with parafilm to prevent evaporation and incubate with gentle shaking for 1 hour at room temperature.

Note: Polyglutamate is an acidic molecule. This post-coating step is necessary in order to saturate the polylysine charges not bound by DNA.

5. Discard the polyglutamate. Add 150 µl TBS per well and then empty the plate, tapping it on a paper towel. Repeat this step twice.

Washing step

6. Add the PBS - 3% BSA - FCS 5% solution to the coated wells and to the uncoated control wells (100 µl per well). Incubate for one hour with gentle shaking at room temperature.

Saturation of the plate

Note: The saturation solution may be prepared in advance, aliquoted and stored at -20°C.

7. Discard the saturation solution. Washing is not required. Add the samples (i.e. human sera diluted 1:300 in PBS - 1.5% BSA - 2.5% FCS - Tween 20 0.05%) in duplicate to the coated and uncoated wells (50 µl per well). Include a positive and a negative control and a blank (i.e. a well containing diluting buffer but no sample). Incubate for 3 hours with gentle shaking at room temperature.

Delivering of the samples

Note: The appropriate sample dilution must be determined beforehand.

Note: It may be helpful, especially when processing a large number of samples, to record the distribution of the samples on a printed grid corresponding to the plate.

8. Discard the samples. Add PBS - Tween 20 1% (150 µl per well) and discard. Then add 150 µl PBS per well. Empty the plate, tapping it on a paper towel. Repeat the washing step with PBS (150 µl) twice.

Washing step

9. Add the detection antibody (in this case, anti-human IgG labeled with alkaline phosphatase) at the appropriate dilution in PBS - 1.5% BSA - 2.5% FCS - Tween 20 0.05% (50 µl per well). Incubate for 3 hours, with gentle shaking, at room temperature or overnight at 4°C.

Adding the labelled antibody

Note: The appropriate dilution of the labeled detection antibody should be prepared beforehand. The manufacturer's recommendations may serve as general guidelines, but the optimal dilution must be optimized for each batch.

Washing step   **10.** Discard the labeled antibody. Add 150 μl PBS-Tween 20 1% per well and then discard. Add 150 μl PBS per well and empty the plate, tapping it on a paper towel. Repeat this step twice.

Chromogenic reaction   **11.** Add 80 μl of substrate to each well and let the colour develop, covering the plate with aluminium foil or keeping it in the dark since the reaction products are photosensitive. Read the plate in a spectrophotometer at 405 nm, subtracting the blank value. Subtract the OD values of the samples on the uncoated plates from the corresponding OD values of the samples on the coated wells.

Note: The substrate solution should be prepared fresh just before use. If necessary, it can be warmed before adding it to the plates in order to accelerate the enzymatic reaction.

Subprotocol 3
Purification of DNA

This technique is used to remove proteins contaminating commercial calf thymus DNA and to reprecipitate and concentrate DNA. The purified DNA may be subsequently used as an antigen in ELISA as described in the appropriate protocol.

■ ■ Materials

Equipment

- Microcentrifuge tubes
- Microcentrifuge
- Pipettemen and tips
- Vortex
- Vacuum desiccator or speedvac evaporator
- Chemical hood

Solutions

- Calf thymus DNA
- 25:24:1 phenol/chloroform/isoamyl alcohol solution
- TE 10/1: Tris 10mM, pH 7.6, EDTA 1 mM
- 3 M sodium acetate, pH 5.2
- Ice-cold 100% ethanol (keep at -20°C)
- Ice-cold 70% ethanol (keep at -20°C)

Preparation

- **25:24:1 phenol/chloroform/isoamyl alcohol** (10 ml):
  5 ml phenol, 4.8 ml chloroform, 0.2 ml isoamyl alcohol. **Warning: phenol is highly toxic;** use under a chemical hood.
- **3 M sodium acetate:**
  Dissolve 40.8 g $CH_3COONa*3H_2O$ in 80 ml distilled water, adjust to pH 5.2 with acetic acid and add water to a volume of 100 ml. Autoclave and store at room temperature.
- **EDTA 0.5 M:**
  Weigh out 11.3 g EDTA tetrasodium salt*$4H_2O$, dissolve in 40 ml distilled water, adjust to pH 7.5 with NaOH 1 M and add distilled water to a volume of 50 ml. **Note that:** EDTA does not dissolve if pH is not close to 8.
- **TE 10/1:**
  Dissolve 0.12 g Tris in 80 ml distilled water, adjust to pH 7.6 adding 37% HCl, add 0.2 ml EDTA 0.5 M and adjust to a volume of 100 ml with distilled water. Autoclave and store at room temperature.

▨▨ Procedure

1. Dissolve 10 mg calf thymus DNA in 10 ml TE; vortex gently; leave for 24 hrs at room temperature; dispense 0.5 ml into each microcentrifuge tube.

2. Add 0.5 ml phenol/chloroform/isoamyl alcohol to each microcentrifuge tube.

3. Vortex vigorously for 10 sec.

4. Spin for 15 sec. at room temperature in a microcentrifuge to separate the two phases (the upper aqueous phase will contain the DNA).

5. Carefully remove the top phase using a 200 ml pipettor or a Pasteur pipette and transfer to a new microcentrifuge tube.

6. If a white precipitate is present at the aqueous/organic interface, repeat steps 2 to 5.

7. Add 0.05 ml of 3 M sodium acetate, pH 5.2, to the DNA solution in each microcentrifuge tube. Mix by vortexing briefly.

8. Add 1 ml of ice-cold 100% ethanol to each tube Mix gently by inverting several times and place on crushed dry ice for 5 min or longer, or overnight in a freezer (-70°C).

9. Spin for 5 min at 900 g in a fixed-angle microcentrifuge.

10. Aspirate off the ethanol supernatant with a Pasteur pipette or a pipettor.

11. Add 1 ml of 70% ethanol, mix gently by inverting the tube several times and spin as in step 9.

12. Aspirate off the supernatant as before.

13. Dry the pellet in a desiccator under vacuum or in a speedvac evaporator (Savant). Alternatively, the tubes can be left to air-dry in a clean and protected environment.

14. Dissolve the dry pellet in 4 ml of sterile distilled water or sterile TE. For long-term storage, dissolve in TE buffer.

15. Read the absorbance of the DNA solution at 280 and 260 nm, diluting 1:100 in distilled water (10 µl in 1 ml water). If low values are obtained, a further extraction and precipitation step will be necessary.

Note: A solution of DNA at 50 µg/ml will yield an optical density of 1 at 260 nm. The ratio of the optical densities at 260 and 280 nm indicates the purity of the sample. For a pure DNA solution, the OD_{260}/OD_{280} is 1.8-2.

Note: Lower values indicate that the sample is contaminated with proteins or phenol.

References

Porstmann T and Kiessig ST. Enzyme immunoassay techniques. An overview. J Immunol Methods 1992; 150: 5-21.

Vogt RF, Phillips DL, Henderson LO, Whitfield W, Spierto FW. Quantitative differences among various proteins as blocking agents for ELISA microtiter plates. J Immunol Methods 1987; 101: 43-50.

Esser P. Edge effect in microwell ELISA. Nunc Bullettin, n.1, October 1985. pp. 1-4. Avrameas S. Amplification systems in immunoenzymatic techniques. J Immunol Methods 1992; 150: 23-32.

Abbreviations

dsDNA	double-stranded DNA
ssDNA	single-stranded DNA
TBS	Tris buffered saline
BSA	bovine serum albumin
mBSA	methyl-BSA
FCS	fetal calf serum
poly-Lys	poly-lysine
poly-Glu	poly-glutamic acid

Chapter 3

Immunoblotting

LAURA CAPONI AND PAOLA MIGLIORINI

Introduction

Immunoblotting is a widely used procedure to analyse antigens
which have been separated and transferred to a membrane.
When applied to protein antigens, it is called Western blotting.
Different techniques can be used to separate the proteins contained in a sample, such as SDS-PAGE (sodium dodecyl sulphate
polyacrylamide electrophoresis, see Appendix 2), which is based
on differences in the molecular weight of proteins, or isoelectro-
focusing, which is based on differences in their isoelectric points.
The separated antigens can then be transferred from the gel to a
membrane for immunological characterisation.

Standard separation procedures can be found described in
most laboratory manuals. Here we will describe the subsequent
steps of the immunoblot procedure, including transfer of proteins from the separation gel to the immunoblot membrane, saturation of the unbound areas of the membrane, incubation with
recognition antibodies and then with a detection reagent, and
finally the detection step. Variations in this procedure required
by particular conditions will be discussed. The dot blot, in which
antigens are placed directly on the membrane, will be examined
in a separate section.

Laura Caponi, University of Pisa, Department of Internal Medicine, Clinical
Immunology Unit, Pisa, Italy
Paola Migliorini, University of Pisa, Department of Internal Medicine,
Clinical Immunology Unit, Pisa, Italy

Blotting

If the proteins have been previously treated with a reagent such as SDS to give them a negative charge, their transfer from the gel to the membrane can be carried out by means of an electric field. Otherwise passive diffusion can be used, although this is a much slower and less efficient procedure.

Generally nitrocellulose membranes are used for immunoblotting, the proteins attaching themselves by means of hydrophobic bonds. Although the binding capacity of nitrocellulose is only moderate compared to that of other materials, it is sufficient for most purposes. If the membrane must be re-probed several times, however, another type of membrane, or nitrocellulose that has been chemically modified so that it can form covalent bonds with the blotted antigens, should be used to avoid antigens being eluted.

Polyvinylidene difluoride (PVDF) membranes, for example, offer a high binding capacity for proteins and are quite resistant to mechanical and chemical stress, more than nitrocellulose. They are frequently used for the blotting of proteins that must be sequenced or subjected to chemical analysis. Nylon membranes also have a high binding capacity for proteins. Furthermore, they are positively charged and thus permit the stable binding of proteins separated by SDS electrophoresis. However, they are not without drawbacks since their complete saturation is difficult to achieve and they exhibit serious background problems when developed with precipitating substrates. For these reasons nylon membranes are used only with chemiluminescence detection systems.

In the wet electrotransfer procedure, the membrane and gel are coupled and placed in a tank filled with buffer and provided with electrodes. An applied electric field results in the transfer of the proteins from the gel to the membrane. After moistening the membrane with buffer, the gel and membrane are assembled in a sandwich between two sheets of filter paper and two porous pads (Fig. 1), the whole assemblage held securely together by clamps in a plexiglass unit. Care must be taken to eliminate any air bubbles between the gel and membrane. The unit can then be placed in the electrotransfer tank, which has been filled with transfer buffer, making sure that the membrane is facing the positive electrode with (Fig. 2).

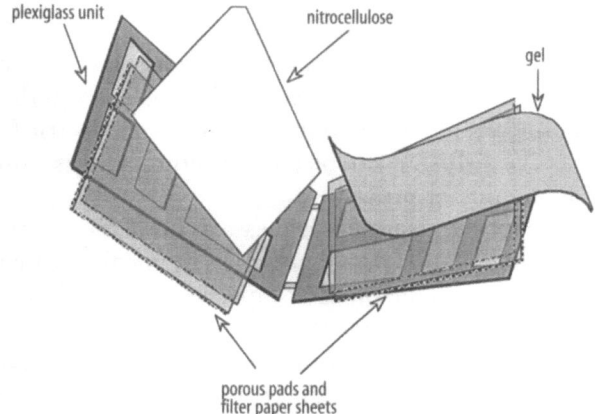

Fig. 1. Assembling the transfer unit.

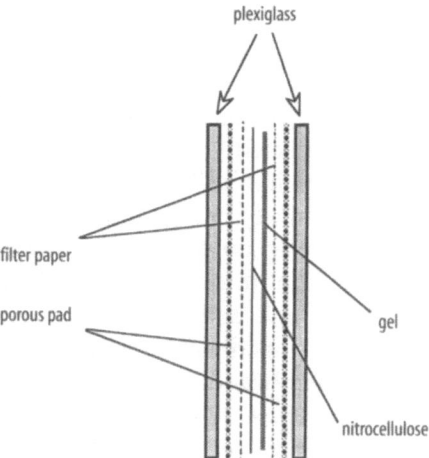

Fig. 2. Cross-section of the assembled electrotransfer unit showing the various layers.

Generally, a Tris-glycine-methanol-SDS buffer is used to transfer the separated proteins to the nitrocellulose membrane (see protocols for details). At high concentrations, however, SDS actually seems to inhibit the binding of proteins to nitrocellulose, and therefore in many laboratories this component is omitted. Without the addition of SDS, however, the gel may swell. Methanol helps to counteract swelling, but can diminish the transfer efficiency, since it also tends to fix the protein on the gel. There-fore, the composition of the transfer buffer represents a compro-mise between different requirements. It must be kept in mind that a transfer buffer containing no SDS will allow a partial re-folding of the protein.

The electrotransfer can be conducted at a constant voltage with a moderate current to induce the transfer. Under these con-ditions, resistance may be expected to drop to a minimum dur-ing the transfer since the ions eluted from the gel will decrease resistance. It is therefore important to use a cooling unit; other-wise the transfer buffer and the gel/membrane unit will heat up, especially if the transfer is performed at high voltage.

The speed of the transfer will depend on the thickness of the gel; generally at least one hour is required, while with a thicker gel even more time must be allowed for an optimal transfer. By low-ering the voltage (and therefore the current), blotting can be car-ried out overnight. The size of the proteins to be transferred re-presents another significant variable; proteins of higher molecu-lar weight will need a longer transfer time. Moreover, the optimal transfer time for large proteins may be too long for the smaller ones, which will then cross the membrane. In such cases it is ad-visable to use nitrocellulose membranes with a small pore size (0.1 - 0.2 μm instead of the more commonly used 0.45 μm) in order to trap the smaller proteins.

Where electrotransfer is not feasible, wet transfer by passive diffusion can be carried out. With this technique, since there is no electric field, the proteins will diffuse out of the gel in all di-rections. This means that it is possible to obtain two replicas of the separated proteins by applying two membrane sheets, one to each side of the gel. The assemblage in this case is placed flat in a tank, weighted down, and covered with transfer buffer. Diffusion time will vary from 36 to 72 hours and the transfer buffer will usually also include EDTA, NaN_3 and beta-mercaptoethanol. The efficiency of transfer is estimated to be between 50 and 70% of that of electric transfer.

Equipment for the transfer of proteins under semi-dry con-ditions is available. The primary advantage of such a procedure is that, since a reduced amount of transfer buffer is needed, notable savings in terms of reagents may be had. It also does not require a cooling system, but the transfer of high molecular weight pro-teins is not as efficient as by the wet electrotransfer technique.

Staining of proteins on the membrane

Whichever method is used, it is generally a good idea to check the efficiency of the transfer before going on to the next step. For proteins, Ponceau S staining can be used; this procedure is simple and the solution can be re-used. Although its sensitivity is not as high as that of Coomassie blue, it is usually sufficient for a quick test. After 10 to 20 minutes of incubation in Ponceau solution, the immunoblot membrane is rinsed with distilled water and the blotted proteins become visible. Rinsing with a buffered saline solution (e.g. TBS or PBS) destains the membrane without interfering with the subsequent steps.

If no protein appears (and you are sure that your staining solution is fresh), it is possible that: (1) the transfer was unsuccessful and the proteins remained on the gel, in which case staining the gel with Coomassie will reveal the proteins; or (2) the amount of proteins is too low to be revealed with Ponceau S, in which case staining a small strip of the membrane with a more sensitive solution (e.g. Coomassie blue) could reveal the proteins. If so, the immunoblotting procedure can still be carried out, since it is a more sensitive detection system than Ponceau staining.

Blocking the membrane

Once stained, the membrane can be cut into strips, each bearing a separate combination of proteins to be probed with different antibody-containing samples (Fig. 3).

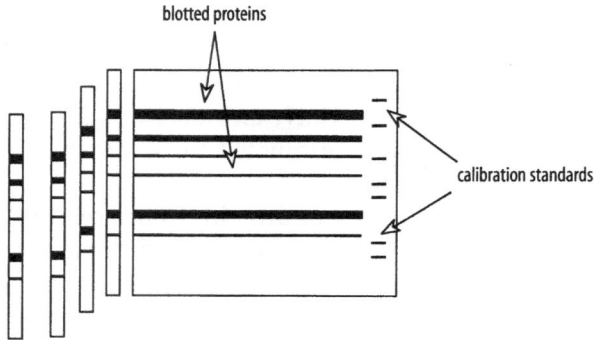

Fig. 3. Cutting strips from a blotted nitrocellulose sheet.

The membranes have a great affinity for proteins, and blocking the unbound sites with non-reactive proteins is necessary to prevent non-specific binding of antibodies. Usually the samples are diluted in the same solution as that used to block the reaction. Many substances can be used for the blocking step. For example, 5% dry non-fat milk dissolved in a buffered solution at physiological pH gives good results; moreover the lyophilised powder is inexpensive and easy to dissolve. Comparable results can be obtained with 2% casein, although in this case the blocking solution must be prepared in advance since casein dissolves very slowly. With warming, it can be prepared in large amounts and then frozen in aliquots to be thawed when needed. Unlike milk solutions, which may undergo significant deterioration, casein solutions can be frozen and thawed several times. This allows the storage and re-use of diluted samples. Other blocking agents such as 1% to 3% gelatine, 5% fetal bovine serum (FCS) or 3% bovine serum albumin (BSA), either alone or in combination, may be used in specific cases but are often accompanied by a higher background.

Incubation of the strips in the blocking solution for 30 to 60 minutes at room temperature is usually sufficient to obtain good saturation.

Sample dilution

The sample containing the antibodies can be diluted in the same solution as that used for blocking. The addition of a mild detergent can help to prevent further non-specific binding, but may remove proteins from the membrane. The sample may consist of whole serum, a purified antibody or a monoclonal antibody, and the amount of dilution must be adjusted accordingly. The optimal serum dilution depends on the antibody titer and in most cases will range from 1:100 to 1:1000. When the antibodies can be quantified, a concentration of 25 µg/ml is usually appropriate.

The sample volume will depend on the membrane surface area and on the type of container used for incubation (tube, incubation tray or sealed plastic bag). The samples must be incubated with the membranes for 2 to 4 hours at room temperature, or for a longer period of time at 4°C, on a rocking platform. Before adding the labelled reagent a washing step must be performed (see below).

Labeled reagents

The antibody bound to the antigen on the strip is usually not directly labelled, and can be revealed in one of two ways: (1) by labelled protein A or protein G, or (2) by a labelled antibody specific for the first antibody. In the former case only those antibody classes that are recognised by protein A or G will be revealed, however (see Appendix 3). Moreover, these proteins cannot be used if the blocking agent is serum. If, on the other hand, a specific labelled antibody is used, any given Ig class or subtype can be detected.

Radioactive labelling is much less widely used at present for protein immunoblotting since enzyme labelling is safer and sensitive enough for most purposes. When a higher sensitivity is required, chemiluminescence can be used. The enzymes most commonly employed for labelling purposes are peroxidase and alkaline phosphatase.

One standard labelling procedure is based on enzyme-labelled antibodies directed towards the antibodies present in the sample (see Fig. 4a). In the conventional immunoblot the enzyme acts on a substrate, generating a coloured compound that precipitates at the site where the labelled reagent is bound. If the label is peroxidase, the optimal reagent for such colour precipitation is chloronaphthol. Diaminobenzidine is no longer widely used since it has been shown to be cancerogenic. If a peroxidase-labelled reagent is used, NaN_3 must be omitted from all solutions since it inhibits the peroxidase reaction. If the label is alkaline phosphatase, the substrate should contain nitroblue tetrazolium, which will precipitate a purple compound where the enzyme-linked antibody is bound.

The sensitivity of this indirect detection procedure is quite high but can be further enhanced by use of the avidin/biotin system. Biotin is a small organic substance with a high affinity for the avidin protein. The detection antibody may be biotinylated rather than enzyme-linked, and avidin followed by the biotinylated enzyme can then be used (see Fig. 4b). The advantages of the avidin/biotin system are its high affinity (greater than that of antigen/antibody complexes) and the fact that one avidin molecule can bind four copies of biotin. The disadvantage of the system is that it requires more steps to be added to the immunoblot procedure.

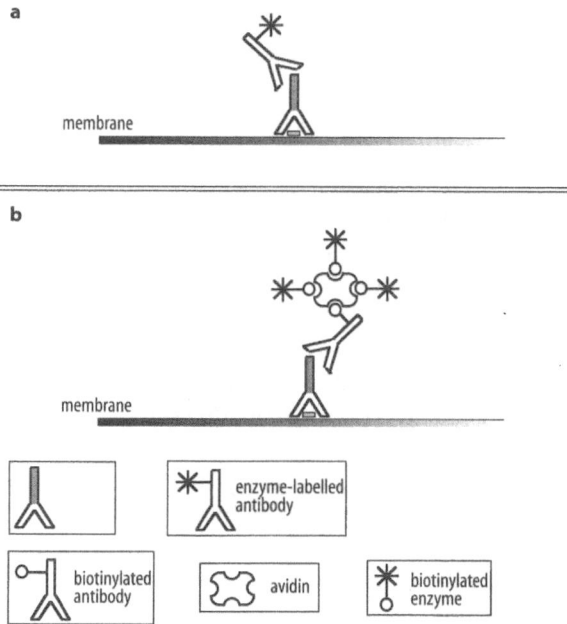

Fig. 4. Immunoblotting detection:
a) the antibody in the sample is detected by an enzyme-labeled secondary antibody;
b) the antibody in the sample is detected by a biotinylated secondary antibody followed by avidin and biotinylated enzyme.

The development step must be carried out in the dark, and the developed membranes must be rinsed in distilled water and stored dry in a light-proof container.

The sensitivity of the immunoblot procedure can be enhanced by the use of chemiluminescent substrates. With this technique the enzyme catalyses a reaction which produces a light emission that can be captured on X-ray or photographic film. Very low amounts of antigen can be detected in this way, and the exposure of the film can be adjusted to obtain optimum results.

The labelled reagents may be diluted in the same medium as that used for the samples. The incubation period at room temperature with gentle shaking will range from 2 to 4 hours, although the samples can also be incubated overnight at 4°C. The incubation time for avidin is much shorter (1 hour).

A control strip incubated with the labelled reagent alone should be used to evaluate the background caused by the binding of the labelled reagent to the blotted proteins.

Washing steps

In the immunoblot procedure, the washing steps carried out before the addition of the labelled reagents and the substrate are of fundamental importance as they help to ensure a low background. Usually the washing solution will be the same as that used in the saturation step and for the dilution of the antibodies. No detergents should be added because this may cause the proteins to be eluted. The washings must be repeated and extensive: at least three washings, each one lasting for five minutes, are required. Sometimes a longer washing time or a higher number of washings can help to diminish the background even further. If serious background problems persist even when using an optimal saturation agent, one might try adding a small amount of a mild detergent such as Tween 20 (0.05% - 0.1%) to the washing solution.

The strips exposed to different samples can be washed together, but it must be kept in mind that when the samples contain a high quantity of antibodies, the antibodies on one sample, if not firmly bound to their target, may "jump" to another strip, particularly during the first washing in the washing procedure. Therefore it is advisable to carry out at least the first of the three washings in separate tubes.

After washing them three times with the saturating solution, the strips should be washed one more time with buffer alone in order to remove the excess of saturating agent before adding the substrate. A final quick rinse with substrate buffer should then be carried out to equilibrate the strips with the pH of the development step.

Dot blot

The dot blot procedure is similar to that of the immunoblot, but the antigens are instead applied - undiluted or combined as a mixture - directly onto the membrane. Specifically, antigen solution (1-2 µl) is carefully dotted onto a sheet or strip of nitrocellulose, and the membrane is then air-dried, blocked and incubated with the various probes.

The dot blot can be used to perform qualitative tests, i.e. for the rapid detection of immunological reactivity in a sample against multiple antigens, each one layered on a spot; if the membrane is too large to fit in a tube, it may be incubated in a swallow dish. Alternatively, many samples can be tested simultaneously using a plate pierced with holes through which the samples may be applied to the membrane, the plate being tightly clamped to the sheet to prevent leakage.

The dot blot can also be used for semiquantitative detections by applying serial dilutions of the antigen or sample.

Protocol
Immunoblotting of ribosomal proteins

Materials

Equipment

- Tank for wet electroblotting
- Power supply
- Nitrocellulose: 0.45 µm pore size is suitable for most purposes
- Tray or tubes to hold strips
- Containers for washing and chromogenic reaction steps
- Pipettemen and tips
- Cutting tool: we use a paper cutter

Reagents

- Salts: Tris, Na_2HPO_4, KH_2PO_4, NaCl, KCl, $MgCl_2$, glycine
- Methanol
- HCl 37%
- Dry non-fat milk: the powdered milk available in grocery stores is suitable if appropriately defatted.
- Labeled detection antibody: goat anti-human IgG (e.g. Sigma).

- Ponceau S
- Dimethylformamide (DMF)
- Trichloroacetic acid (TCA)
- 5-Bromo-4-chloro-3-indolyl phosphate (BCIP)
- Nitro blue tetrazolium (NBT)

Solutions

- Blotting buffer
- Ponceau S solution: 0.5% in TCA 3%
- Tris-buffered saline (TBS: Tris 10 mM, NaCl 150 mM)
- Saturating/diluting buffer: TBS - milk 5%
- Tris pH 9: (Tris 0.1 M, NaCl 0.1 M, $MgCl_2$ 5 mM)
- Stock solutions of BCIP and NBT

Preparation

- **Blotting buffer:**
  Dissolve 6 g Tris, 28.75 g glycine in one liter of water. Add 400 ml methanol and distilled water to a final volume of 2000 ml. Store at 4°C.

Note: The membrane must be pre-equilibrated with blotting buffer for at least 15 minutes prior to the electroblotting step.

- **TCA 3%:**
  Dissolve 3 g TCA in 100 ml distilled water.
- **Ponceau solution:**
  Dissolve 0.5 g Ponceau S in TCA 3%. The solution can be reused. Store at room temperature.
- **TBS:**
  Dissolve 1.21 g TRIS and 8.76 g NaCl in 800 ml distilled water, adjust to pH 7.3 with HCl 37%, and add distilled water to a final volume of 1000 ml.
- After the electrotransfer is completed, prepare a suitable amount of **saturating/diluting solution:**
  TBS–non fat milk 5%: add 12.5 g dry milk to 250 ml TBS.
- **Tris pH 9:**
  Dissolve 12.11 g Tris, 5.844 g NaCl and 0.476 g $MgCl_2$ in 800 ml water, adjust to pH 9 with HCl 37%, and add distilled water to a final volume of 1000 ml.

– **Stock solution of BCIP:**
  Add 50 mg BCIP in 1 ml of 100% DMF. Store at -20°C.
– **Stock solution of NBT:**
  Add 75 mg NBT in 1 ml of 70% DMF. Store at -20°C.
– **Substrate solution:**
  Add 33 µl BCIP stock solution and 44 µl NBT stock solution to
  10 ml Tris, pH 9. Prepare just before the last washing step.

Procedure

1. Lay a porous pad and a sheet of filter paper on each side of the **Blotting of the** open immunoblotting unit. Place the gel in the middle of the **proteins** filter paper on one side.

Note: Cut off one small corner of the NC sheet to mark its orientation. Pre-equilibrate the nitrocellulose in blotting buffer for 10-15 minutes in a clean container. Dry nitrocellulose (NC) is fragile and must be handled with care. Always use gloves to avoid contamination.

2. Lay the NC membrane very carefully over the gel. Lay a sheet of filter paper on the NC membrane and add a porous pad. Close the sandwich.

Note: Make sure that no air bubbles are trapped between the layers, in particular between the gel and the NC. To eliminate bubbles roll a glass rod over the layers before tightly closing the sandwich, or alternatively assemble the sandwich in transfer buffer.

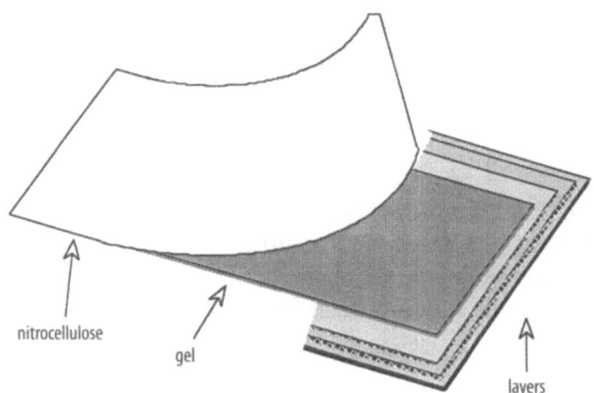

Fig. 5.

3. Place the sandwich in the tank and fill the tank with transfer buffer. Place the tank cover, and turn on the power supply and the cooling unit. Raise the voltage to 300-400 mAmp and let the electrotransfer run for 2 hours.

4. Turn off the power supply and the cooling unit. Extract the sandwich, remove the membrane and let it drain.

Note: Drying nitrocellulose will stick to almost any surface. If the NC has adhered, you will have to wet it again to move it. If necessary, evaluate the efficacy of the transfer by staining the gel with Coomassie.

Staining the blot

5. Pour Ponceau solution into a clean plastic or glass tray. Gently place the NC in the tray. Incubate for 10-20 minutes.

6. Extract the membrane and gently rinse it with water until the bands appear. Drain the membrane and let it dry.

Note: The Ponceau solution can be re-used several times

Cut the strips

7. Cut strips containing all of the blotted proteins from the NC sheet. Manipulate them carefully using fine forceps with flattened tips.

Note: See Figure 3 in the text. Every strip should be labeled on the back in ink with an identifying number.

Blocking the matrix

8. Place the strips in a tray containing an appropriate amount of blocking solution. Cover the tray and let it stand at room temperature for 45-60 minutes, occasionally shaking it gently.

Note: The blocking solution may be re-used for the washing steps. Dilute the samples while the nitrocellulose is blocking.

Incubation with probing antibodies

9. Incubate each strip with the corresponding diluted sample for 3-4 hours at room temperature with gentle shaking.

Note: The incubation step can be carried out in tubes slightly longer than the strips, or in commercially available incubation trays. Be sure that the strips are completely covered with sample solution during incubation. Afterwards, the diluted sample may be stored and re-used.

10. Place the strips in a tray containing the washing solution and shake them gently at room temperature for at least 5 minutes. Repeat twice, changing the washing solution each time.

Washing

Note: A thorough washing step is essential to minimise background. Usually all of the strips can be washed together in the same tray, but sometimes contamination of strips by antibodies coming from different samples occurs (see text). Under these circumstances, at least the first of the three washings should be carried out separately for each strip.

11. Incubate the strips with the detection antibody at the optimal dilution. All of the strips may be incubated in a single container if the labeled antibody is the same for all of the samples. Incubate for 3-4 hours at room temperature with gentle shaking.

Incubation with labelled antibody

Note: The appropriate dilution of labelled antibody must be prepared beforehand. The manufacturer's recommendations may serve as general guidelines, but the dilution must be optimized for each batch.

12. Place the strips in a tray containing the washing solution and shake gently at room temperature for at least 5 minutes. Repeat this step twice, changing the washing solution each time. Then wash once with TBS buffer lacking the blocking agent for 5 minutes. Discard the buffer and immerse the strips briefly in a small amount of substrate buffer.

Washing

Note: In addition to the regular washing steps, a final rinse with TBS buffer is required to remove excess blocking agent. The last washing with substrate buffer serves to equilibrate the strips to the alkaline pH necessary for the chromogenic reaction.

13. Add the substrate solution to the tray, making sure that all of the strips are immersed. Place the tray in a light-free environment or cover it completely with aluminium foil. Wait for the reaction to take place; when the coloured bands have appeared, immediately wash the strips in distilled water and let them dry.

Development with substrate

Note: Fresh substrate solution must be prepared each time (and protected from light since the substrate is light-sensitive). The dried strips may be stored in the dark for future reference.

References

Salinovich O and Montelaro RC. Reversible staining and peptide mapping of proteins transferred to nitrocellulose after separation by sodium dodecylsulfate-polyacrylamide gel electrophoresis. Anal Biochem 1986; 156: 341-347.

Gershoni JM. Protein blotting: A manual. Methods Biochem Anal. 1988; 33: 1-58.

Stott DI. Immunoblotting and dot blotting. J Immunological Methods 1989; 119: 153-187.

Abbreviations

NC	nitrocellulose
TBS	Tris-buffered saline
TTBS	Tween 20 - Tris-buffered saline
TCA	trichloroacetic acid
DMF	dimethylformamide
BCIP	5-bromo-4-chloro-3-indolyl phosphate
NBT	nitroblue tetrazolium

Chapter 4

Immunoprecipitation

PAOLA MIGLIORINI

Introduction

Immunoprecipitation is a technique that allows the identification and separation of many antigens using antibodies that bind specifically to them. This procedure is based on the separation of the antibody-bound antigen from unbound antigens by means of reagents specific for the Fc portion of the immunoglobulins, usually protein A or protein G. Different preparations of unsolubilised protein A or G are commercially available. The matrix they are coupled to is usually sepharose, agarose or acrylic resins.

The different steps of immunoprecipitation can be summarised as follows:

- Labelling of the antigen (not always necessary)

- Solubilisation of the antigen

- Pre-clearing

- Precipitation of the antibody-bound antigen

- Detection of the antigen

Labelling of the antigen

Radiolabelling is performed with ^{125}I, with ^{32}P when phosphorylated proteins are being studied, or with $[^{35}S]$methionine if a metabolic label is required. Radioactive (in particular, iodine) labelling is now much less widely used than enzyme labelling due to the introduction of biotin esthers which are admirably suited to the labelling of soluble and cell-surface proteins.

Paola Migliorini, University of Pisa, Department of Internal Medicine, Clinical Immunology Unit, Pisa, Italy

No labelling is necessary if the immunoprecipitated antigen can be traced for an "intrinsic" signal, e.g. enzymatic activity. Soluble proteins can be directly labelled. Cell surface proteins, on the other hand, require a different labelling procedure as well as a solubilisation step.

Solubilisation of the antigen

Surface-labelled cells can be lysed by a detergent-containing buffer. One standard lysis buffer, RIPA, is composed of Tris-HCl 50 mM at pH 7.5, NaCl 150 mM, NP40 1%, sodium deoxycolate 0.5% and SDS 0.1%. Alternatively, Triton X-100 or NP40 alone can be used. To avoid extensive denaturation of the antigen, moderately denaturing agents such as deoxycolate are preferable to SDS. Lysis buffers usually contain a cocktail of protease inhibitors, preferably ones with different mechanisms of action, such as phenylmethylsulfonyl fluoride (PMSF), pepstatin or leupeptin, and aprotinin to block proteolytic degradation of the solubilised proteins.

Lysis is usually carried out for 20 to 30 minutes on ice. The cell/lysis buffer ratio can vary and 1 ml of buffer may be used to lyse 10^5-10^7 cells. Low speed centrifugation permits the recovery of the solubilised antigen from the supernatant.

If a detergent cannot be used, the cells can be mechanically ruptured by sonication or by Potter or Dounce homogenisers. Sucrose 0.25 M - 0.33 M may be added to help preserve the integrity of the organelles during rupturing of the cell membranes.

It is possible to lyse cells in suspension by osmotic shock using hypotonic buffers. This often does not damage the cells sufficiently to result in the release of the intracellular antigen, however, in which case various amounts of detergent may be added to the buffer.

Pre-clearing

The antigen mixture can be pre-cleared to enhance the specificity of immunoprecipitation. Indeed, antigens can bind either directly to protein A or G or to the matrix that protein A or G is coupled to, or interact with antibodies outside the Fab region.

Different procedures can be used for the pre-clearing step. The antigen mixture can be incubated on the same protein A or G preparation as that used for immunoprecipitation, or on a less expensive and less highly purified insoluble protein A or G preparation (Fig. 1) such as SAC (fraction I of the Cowan strain of Staphylococcus or Pansorbin (heat-killed Cowan Staphylococcus, Calbiochem). When monoclonal antibodies are used for immunoprecipitation, pre-clearing can be performed on an irrelevant isotype-matched antibody coupled to protein A or G. If a whole immune serum is used, a pre-immune serum is the reagent of choice for the pre-clearing step (Fig. 2).

Each absorption step is usually carried out for 1 to 2 hours at low temperatures. The number of pre-clearing steps and the amount of protein A to be used will depend on the experimental conditions: a relatively pure antigen preparation and a high affinity antibody/antiserum will require shorter pre-clearing periods and lower amounts of protein A.

Precipitation of antibody-bound antigen

Antigen-antibody binding can either take place in the liquid phase with the immune complex subsequently binding the insoluble protein A, or alternatively it can form on the solid phase if the antigen reacts with the protein A- or protein G-bound antibody.

The addition of protein A-bound antibodies to the antigen mixture is preferable when whole antiserum is used for immunoprecipitation. In this case the appropriate amount of antiserum is incubated on insoluble protein A for 30 to 60 minutes; the

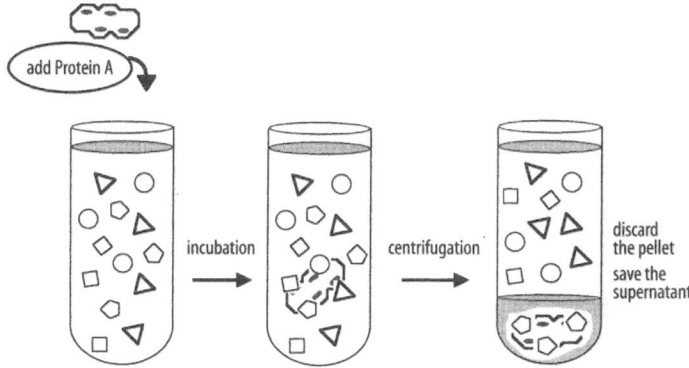

Fig. 1. Preclearing with protein A

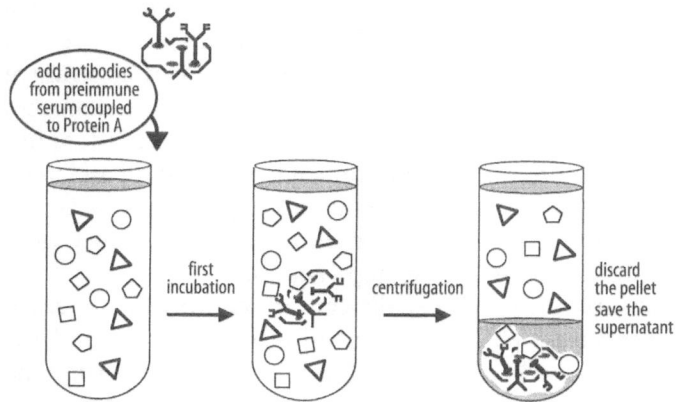

Fig. 2. Preclearing with antibodies from preimmune serum

unbound proteins are then washed off and the protein A-antibody complex is added to the pre-cleared antigenic mixture.

Alternatively, polyclonal or monoclonal antibodies can be added to the antigen mixture, which is then incubated for 1 to 2 hours at room temperature or overnight at 4°C. Insoluble protein A or G is added and the incubation is carried out for 30 to 60 minutes (Fig. 3). This procedure must be modified if the antigen-specific antibody binds poorly to protein A or G (see Appendix 3), if the affinity of the antibodies is low, or if a monoclonal antibody that binds to a non-repetitive antigenic determinant is used. In such cases a second antibody specific for the Fc portion of the antigen-specific antibody allows the formation of a stable immune complex.

The insoluble immune complexes are recovered by low speed centrifugation and the unbound antigens are washed off. Lysis buffer is generally used for the washing steps, although to further increase the specificity of immunoprecipitation high salt buffers (e.g. Tris 10 mM, NaCl 500 mM, Triton 1%) can also be employed.

Detection of the antigen

In a preparatory immunoprecipitation procedure, the antigen can be recovered by subjecting the matrix to low pH treatment

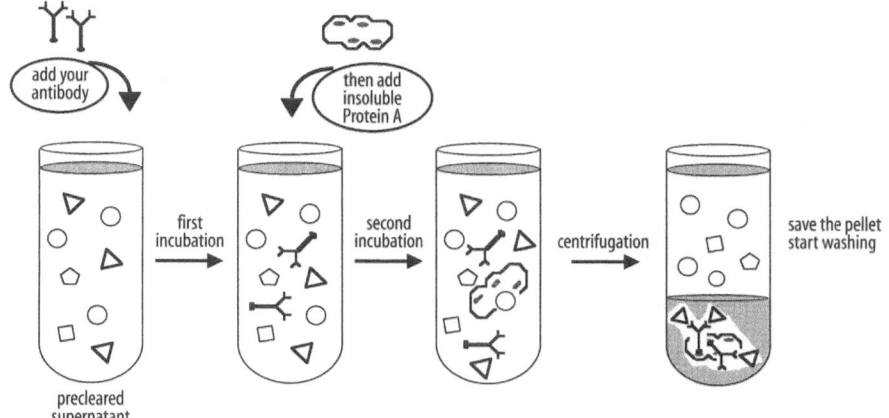

Fig. 3. Immunoprecipitation procedure

and then isolating the antigen by chromatography (size exclusion or ionic). However, immunoprecipitation is usually carried out as an analytical procedure, in which case the immune complex, coupled to unsolubilised protein A or G, must be resuspended in sample buffer, boiled, centrifuged, and the supernatant loaded onto a polyacrylamide gel.

If the antigen is radioactively labelled, the gel is dried and then subjected to autoradiography. If the antigen is biotin-labelled, the gel is transferred to nitrocellulose, probed with labelled avidin, and developed with a precipitating substrate or by enhanced chemiluminescence.

Protocol
Immunoprecipitation of labelled antigens

Materials

Equipment

- Pipettemen and tips
- Eppendorf tubes (1.5 ml)
- Cell scrapers

- Rocking platform
- Centrifuge
- pHmeter
- Pasteur pipettes
- Vacuum pump
- 3 large trays
- Tweezers (small and large)
- Scissors
- Autoradiography film
- Film cassette

Reagents and solutions

Labeling:
- Sulfosuccinimidyl-6'-(biotinamido)-6-hexanamido hexano-ate
- RPMI 1640
- Dulbecco's phosphate-buffered saline (DPBS: KCl 2.7 mM; KH_2PO_4 1.5 mM; NaCl 137 mM; Na_2HPO_4 8.5 mM)
- Trypsin-EDTA 10x (cell culture tested)
- Sodium bicarbonate buffer ($NaHCO_3$ 50 mM, pH 7.5 - 8.5)
- Phosphate-buffered saline (PBS, pH 8: Na_2HPO_4 50 mM; NaCl 100 mM; KH_2PO_4 10 mM, pH 8)
- Phosphate-buffered saline (PBS, pH 7.3: Na_2HPO_4 50 mM; NaCl 100 mM; KH_2PO_4 10 mM).
- Washing buffer (Tris 10 mM; NaCl 140 mM, pH 8).
- Lysis buffer containing protease inhibitors (Tris 10 mM; NaCl 150 mM; Triton X-100 1%; EDTA 1 mM; PMSF 1 mM; pepstatin 1 µM; leupeptin 10 µM)

Immunoprecipitation:
- Sepharose-protein A or Sepharose-protein G
- TETN 250 (Tris-HCl 25 mM, pH 7.5; EDTA 5 mM, pH 7.5; NaCl 250 mM; Triton X-100 1%)
- Bovine serum albumin (BSA) 5% in TETN 250
- TTN 150 (Tris-HCl 10 mM, pH 8; NaCl 150 mM; Triton X-100 1%)
- Sample buffer 4x
- TTN 500 (Tris-HCl 10 mM, pH 8; NaCl 500 mM; Triton X-100 1%)
- Ponceau Red solution

Enhanced Chemoluminescence
- Tris-buffered saline (TBS: Tris 10 mM; NaCl 150 mM)
- Tris-buffered saline Tween 20 0.1% (TTBS: Tris 10 mM; NaCl 150 mM; Tween 20 0.1%)
- HRP-labelled Neutravidin
- ECL Western blot detection reagents (solution A and solution B)
- Developer/replenisher liquid (**note: highly toxic**)
- Fixer/replenisher liquid (**note: highly toxic**)

Preparation

- **Sodium bicarbonate buffer:**
  Dissolve 4.2 g $NaHCO_3$ in 900 ml distilled water; adjust to pH 8 adding HCl 37% dropwise, add distilled water to a volume of 1000 ml.
- **PBS pH 7.3:**
  Dissolve 1.42 g Na_2HPO_4, 8.76 g NaCl and 13.6 g KH_2PO_4 in 900 ml distilled water, adjust to pH 7.3 with HCl 37%, add distilled water to a final volume of 1000 ml.
- **PBS pH 8:**
  Dissolve 1.42 g Na_2HPO_4, 8.76 g NaCl and 13.6 g KH_2PO_4 in 900 ml distilled water, adjust to pH 7.5-8.5 with HCl 37%, add distilled water to a final volume of 1000 ml.
- **Dulbecco's PBS:**
  Dissolve 0.2 g KCl, 0.2 g KH_2PO_4, 8.0 g NaCl and 1.14 g Na_2HPO_4 in 900 ml distilled water, adjust to pH 7.4 with HCl 37%, add distilled water to a final volume of 1000 ml.
- **Washing buffer:**
  Dissolve 1.21 g Tris and 8.2 g NaCl in 900 ml distilled water, adjust to pH 8 with HCl 37%, add distilled water to a volume of 1000 ml.

Stock solutions for lysis buffer:
1. **Tris for lysis buffer:**
   Dissolve 1.21 g Tris and 8.7 g NaCl in 900 ml distilled water, adjust to pH 8 with HCl 37%, add distilled water to a volume of 1000 ml.
2. **EDTA 0.5 M:**
   Weigh out 11.3 g EDTA tetrasodium salt*$4H_2O$, dissolve in 40 ml distilled water, adjust to pH 7.5 with NaOH 1 M, and add

distilled water to a final volume of 50 ml. **Warning: EDTA will not dissolve if the pH is not close to 8.**

3. **PMSF 0.2 M:**
   Dissolve 1.74 g PMSF in 50 ml absolute ethanol. **Warning: PMSF is extremely toxic. Wear gloves and a mask when handling the powder.**

4. **Pepstatin 1 mM:**
   Dissolve 0.7 g pepstatin in 1 ml absolute ethanol. Pepstatin should be used at a concentration of 1 μM.

5. **Leupeptin:**
   Dissolve 0.5 g leupeptin in 1 ml distilled water. Leupeptin should be used at a concentration ranging from 1 to 100 μM, usually 10 μM.

6. **LYSIS BUFFER:**
   Combine 49.05 ml Tris for lysis buffer, 0.5 ml Triton X-100, 0.1 ml EDTA 0.5 M, 0.25 ml PMSF, 0.05 ml pepstatin, and 0.05 ml leupeptin.

– **TETN 250:**
  Dissolve 3.03 g Tris and 14.61 g NaCl in 800 ml distilled water, and adjust to pH 7.5 with HCl 37%. Add 10 ml EDTA stock solution, 10 ml Triton X-100, and distilled water to a final volume of 1000 ml.

– **TTN 150:**
  Dissolve 1.21 g Tris and 8.76 g NaCl in 800 ml distilled water, and adjust to pH 7.5 using HCl 37%. Add 10 ml Triton X-100 and then add distilled water to a volume of 1000 ml.

– **TTN 500:**
  Dissolve 1.21 g Tris and 29.22 g NaCl in 800 ml distilled water, and adjust to pH 7.5 using HCl 37%. Add 10 ml Triton X-100 and then distilled water to a volume of 1000 ml.

– **Sample buffer 4x:**
  Dissolve 400 mg SDS in 2.5 ml Tris 0.5M, pH 6.8. Add 2.3 ml glycerol and 1 mg Bromophenol blue. Store at room temperature. When the sample has to be reduced, prepare a small amount of reduced sample buffer (1 volume of β-mercaptoethanol + 4 volumes of sample buffer 4x) and discard the leftover.

- **TBS:**
  Dissolve 1.21 g Tris and 8.76 g NaCl in 800 ml distilled water, adjust to pH 7.3 with HCl 37%, and add distilled water to a final volume of 1000 ml.
- **TTBS:**
  Dissolve 1.21 g Tris and 8.76 g NaCl in 800 ml distilled water, adjust to pH 7.3 with HCl 37%, add 1 ml Tween 20, and then add distilled water to a final volume of 1000 ml.
- **Ponceau Red solution:**
  Dissolve 0.5 g Ponceau Red powder in 100 ml of 3% trichloroacetic acid (prepared by adding 3 g CCl_3COOH to 100 ml distilled water).

Procedure

1. Dissolve 2 mg IgG in 1 ml of 50 mM sodium bicarbonate buffer, pH 7.5 - 8.5. Immediately prior to use, dissolve 1 mg sulfosuccinimidyl-6'-(biotinamido)-6-hexanamido hexanoate in 1 ml water. Add 75 μl of the biotin solution to the IgG solution. Incubate at room temperature for 30 min.

   Labelling of soluble antigen

Note: Soluble proteins can be labelled at a concentration ranging between 0.5 and 10 mg/ml. Lyophilized proteins can be directly dissolved in 50 mM sodium bicarbonate buffer (pH 7.5 - 8.5).

2. Dialyse extensively against PBS to remove unreacted biotin. The IgG solution may be stored at 4°C for a few days or at −20°C in small aliquots for longer periods. Avoid repeated freezings/thawings.

Note: Protein solutions in buffers other than bicarbonate should be dialysed against 50 mM sodium bicarbonate buffer (pH 7.5 - 8.5), or other buffers such as PBS (pH 7.5 - 8.5).

3a. If the cells are grown in suspension, collect the culture medium containing the cells in one or more tubes.
   If the cells grow adherent to the plastic, they can be detached either mechanically by scraping (follow 3a.1) or chemically (follow 3a.2) using trypsin-EDTA solution.

   Labelling of cell surface antigen, a) Cells in suspension

3a.1. Scraping: Aspirate the culture medium from the flasks and add 5 ml/flask of Dulbecco's PBS. Detach the cells from the flask by scraping and transfer the 5 ml PBS containing the suspended cells to a tube. Add another 5 ml DPBS to the flask, using it to rinse; add to the tube containing the cell suspension.

3a.2. Detachment by trypsin-EDTA: Dilute trypsin-EDTA 1:10 in DPBS; warm to 37°C and add 5 ml to each flask. Incubate at 37°C for 5 min and then check the number of detached cells under a microscope. If necessary, prolong incubation until most of the cells are in suspension. Transfer the cell suspension to a tube. Add 5 ml RPMI 1640 5% FCS to the flask, using it to rinse; add to the tube with the cell suspension (Note: This washing step is necessary to obtain optimal recovery of the detached cells; fetal calf serum blocks trypsin digestion).

Note: Either of these procedures may cause damage to cell membranes. A partial digestion of cell surface proteins by trypsin can take place, especially in those proteins containing well-exposed trypsin-sensitive sequences.

4a. Centrifuge the cells at 400 g for 5 min. Dry the pellet and resuspend by gently pipetting into 5 ml RPMI. Repeat this centrifugation step twice with RPMI and once with ice-cold PBS, pH 8. Dissolve 1 mg sulfosuccinimidyl-6'-(biotinamido)-6-hexanamido hexanoate in 1 ml PBS, pH 8.

5a. Re-suspend the cells (10 x 10^6) in 1 ml biotin solution and incubate for 15 min at room temperature, swirling the tube every 5 min. Protect from light.

Note: A lower number of cells may also be labelled, reducing proportionally the volume of biotin solution required.

6a. Centrifuge at 400 g for 5 min. Add 1 ml of cold washing buffer and centrifuge at 400 g for 5 min. Dry the pellet and perform one more washing step.

7a. Add 1 ml of cold lysis buffer containing protease inhibitors to the cell pellet. Place on ice for 15 min, vortexing for 30 seconds every 5 minutes.

8a. Centrifuge for 10 min at 4°C and 20,000 g. Use this supernatant for the immunoprecipitation assay.

3b. Aspirate the culture medium from 5 flasks (75 cm^2) of confluent cells; add 10 ml RPMI 1640 and aspirate. Repeat this sequence twice.

4b. Add 10 ml of ice-cold PBS, pH 8.
Dissolve 2.5 mg sulfosuccinimidyl-6'-(biotinamido)-6-hexanamido hexanoate in 10 ml PBS, pH 8.

5b. Add 2 ml biotin solution to each flask.
Incubate on a rocking platform for 15 min at room temperature.

6b. Add 3 ml of cold washing buffer to each flask to stop the reaction. Wash twice with cold washing buffer.

7b. Add 2 ml of cold lysis buffer containing protease inhibitors to each flask. Incubate on a rocking platform for 15 min at 4°C.

8b. Centrifuge for 10 min at 4°C and 20,000 g. Use the supernatant for the immunoprecipitation assay.

Note: Less disruption of the cell membranes will occur if the cells are labelled while they are still adherent to the plastic surface, but a lower yield of labelled membrane proteins is to be expected in this case as only a part of the cell surface is accessible to the label. Moreover, the cells may be polarized. Some of the membrane proteins may be expressed mostly or exclusively either on the portion of the cell membrane that is adherent to the substrate or on the portion that is exposed to the culture medium. Therefore, the distribution of the antigen on the cell membrane should be taken into account when choosing the cell labelling method to be used.

9. Weigh out 100 mg protein A-Sepharose.

Pre-clearing

10. Add 5 ml washing buffer and incubate for 20 min at room temperature.

11. Centrifuge at 400 g for 5 min; dry the pellet and repeat step 10 twice.

12. Centrifuge at 400 g for 5 min; resuspend the pellet in 5 ml TETN 250 immunoadsorption buffer containing 50 mg/ml BSA. Incubate for 1 hour at room temperature on a rocking platform. The protein A blocked with BSA will be used here-after.

13. Transfer 500 μl of the protein A suspension to an Eppendorf tube. Centrifuge, aspirate the supernatant, and add the labelled cell lysate to the protein A pellet.

14. Incubate for 2 hours at 4°C on a rocking platform.

15. Centrifuge; transfer the labelled cell lysate to a new protein A pellet. Incubate for 2 hours at 4°C on a rocking platform. Centrifuge and transfer the supernatant to the protein A-NRS pellet generated in step 16.

16. While the cell lysate is incubating in step 14, transfer 500 μl of fresh protein A suspension to an Eppendorf tube labelled "protein A-NRS"; centrifuge and add 100 μl TTN 150 buffer containing 5 μl normal rabbit serum. Incubate for 2 hours at 4°C on a rocking platform. Centrifuge, wash once with 1 ml TTN 150, centrifuge again and dry the pellet.

17. Incubate the labelled cell lysate from step 8 on the protein A-NRS pellet for 2 hours at 4°C on a rocking platform. Centrifuge and use the supernatant for the immunoprecipitation assay.

Note: The supernatant can be stored frozen. Otherwise proceed to step 18.

18. Transfer 500 μl of fresh protein A suspension to each of two Eppendorf tubes (nos. 1 and 2). Centrifuge and add 100 μl TTN 150 buffer to each tube. Add 5 μl normal rabbit serum to tube no. 1 and 5 μl rabbit immune serum to tube no. 2. Incubate for 2 hours at 4°C on a rocking platform.

19. Centrifuge the tubes, wash each once with 1 ml TTN 150; centrifuge and dry the pellets.

20. Add 200 μl of pre-cleared lysate to tube no. 1 and the same amount to tube no. 2.

21. Incubate the tubes overnight at 4°C on a rocking platform.

22. Wash each tube with 1 ml TTN 150 by vortexing. Repeat 5 times.

23. Wash each tube with 1 ml TTN 500 by vortexing. Repeat 3 times.

24. Centrifuge the tubes, dry the pellets, and add 50 μl sample buffer. Vortex each pellet, boil for 3 min, centrifuge and load the supernatant on an acrylamide gel.

Note: The supernatant can be frozen at this step. The number of pre-clearing procedures must be adjusted to the conditions of each experiment.

25. Perform the electrophoresis and blotting procedures (see Appendix 2 and Immunoblotting chapter). Biotinylated MWM can be used to correctly determine the molecular weight of the precipitated antigen.   **Antigen detection**

Note: Nitrocellulose suitable for ECL is recommended to obtain optimal results. Blots can be coloured using Ponceau Red before blocking in order to check the efficiency of the transfer.

26. Non-specific binding sites will be blocked by incubating the membrane in 5% dry non-fat milk dissolved in Tris-buffered saline (TBS) for 30 min at room temperature on a rocking platform.

27. Wash twice in as large a volume of TTBS as possible.

28. During the washing step, dilute the HRP-labelled Neutravidin in TTBS (optimal dilution 1:2000 – 1:10 000).

29. Incubate the membrane in the diluted Neutravidin for 45 – 60 at room temperature.

30. Wash the membrane 5 times (for 5 min each time) in TTBS and once for 5 min in distilled water.

Note: Note: The membrane can be kept in distilled water for an extended period of time before developing.
Fireston GL and Wingut SD. Immunoprecipitation of proteins in Methods in Enzimology vol. 182 pag. 688 – 700, 1990

31. While the membrane is being washed, mix an equal volume of detection solution A with detection solution B. The final volume should be sufficient to cover the membrane.

32. Incubate the membrane for precisely 1 minute in this solution at room temperature without shaking.

33. Drain off the excess detection reagent and place the membrane, protein side up, in the film cassette.

Note: Work as quickly as possible to minimise the time lapse between the incubation step and exposure of the film. Make sure that there is no detection reagent in the film cassette. Wrap the membrane in transparent plastic to keep it from wetting the film.

34. Switch off the lights in the room and carefully place the autoradiography film on top of the membrane; close the cassette and expose for 30 seconds.

35. Remove the film and submerge it in developer/replenisher liquid for 15 - 20 seconds.

36. Wash the film in water for 5 seconds and immerse it in fixer/replenisher liquid for 30 seconds.

Note: The developer/replenisher and fixer/replenisher solutions must be handled in the dark.

37. Wash the film in water for 1 minute and then let air-dry.

38. Repeat steps 35 to 37 if needed, changing the exposure time.

Note: Following the detection of the labelled proteins by ECL, it is possible to probe the membrane with specific enzyme-labelled antibodies, subsequently developing them by means of precipitating substrates. In this case, the following sequence of steps should be carried out:

39. Wash the membrane twice (for 10 min each time) in a large volume of TTBS at room temperature.

40. Block the membrane by immersion for 30 minutes in 5% dry non-fat milk dissolved in TBS.

41. Carry out the immunodetection procedure as described in the immunoblotting chapter.

Chapter 5

Immunofluorescence

MARINA FABBI AND MICAELA TISO

Introduction

Immunofluorescence is a technique which allows the immuno-localisation of targets by means of fluorochrome-conjugated antibodies. Fluorescence occurs when a molecule excited by light of one wavelength returns to the unexcited state by emitting light of a longer wavelength. The exciting and emitted light, being of different wavelengths, can be separated using optical filters. This is a sensitive technique since a positive signal is detected against a negative background. Furthermore, the ability to detect fluorescence simultaneously from two, three or four compounds fluorescing at different wavelengths allows multi-parametric analysis.

A wide variety of fluorescently labelled antibodies are currently available, and the choice of fluorochromes to be used therefore depends on the instrument and its set of optical filters. Care should always be taken, however, to select the combination of fluorochromes which allows both efficient excitation and good optical separation of the two (or more) signals. Table 1 shows some fluorescent molecules and their properties.

The most commonly used fluorochrome for both light microscopy and flow cytometry is fluorescein (FITC). Its absorption maximum is conveniently close to the emission lines from both the mercury lamp (microscope) and the argon ion laser (flow cytometer). Moreover, it exhibits both a high extinction coefficient and quantum efficiency. When a second label is required, the fluorochrome of choice is tetramethylrhodamine (TRITC) if

Marina Fabbi, National Institute for Cancer Research, Genoa, Italy
Micaela Tiso, DIMES, University of Genoa, Genoa, Italy

Table 1. Fluorochromes commonly used to label antibodies

Fluorochrome	Excitation maxima (nm)	Emission-maximum (nm)	Colour
Fluorescein	495	520	Green
Tetramethylrhodamine	543	570	Red
Texas red	596	620	Red
Princeton red	490	570	Red
R-Phycoerythrin	564, 495	576	Orange
Allophycocyanin	650	660	Deep red

samples are analysed by microscopy, and R-phycoerythrin (PE) when samples are analysed by flow cytometry. PE can be efficiently excited by the same laser emission (488 nm) as that required for FITC.

Flow cytometry is a technique based on rapid measurements carried out on cells as they flow in a fluid stream one by one through a sensor point. Its key feature is that measurements are made separately on each particle within the suspension in turn and the individual measurements can be either stored in the computer or processed immediately into histograms and cytograms (dot plots). Analysing cells at a rate of 5000 per second and measuring six parameters for each cell, flow cytometers can generate vast amounts of data. They are thus particularly suitable for the precise measurement of cellular subsets identified by antibodies within a heterogeneous population.

When the immunolocalisation of the target must be correlated with the different subcellular compartments, optical microscopy and, in some cases, confocal microscopy become the techniques of choice.

In general, staining cells with antibodies is a relatively straightforward procedure. The cells are mixed with an appropriately diluted antibody and then incubated to allow antigenic binding to take place. If the antibody is already conjugated to a fluorochrome, the sample can be washed and processed directly. Otherwise it requires further incubations with fluorescent reagents and washing steps before analysis. We will consider protocols for both indirect and dual immunofluorescence techniques, since they represent the most common applications. Protocols for direct and multiple immunofluorescence techniques can be easily extrapolated from these procedures.

Sample preparation

The sample preparation procedure must take into account the instrument being used to observe the specimen. When a flow cytometer is available, the purpose of sample preparation is to obtain a suspension of single cells which will pass freely through the flow system. To generate rapid and accurate results, the sample should contain as little debris as possible. Tissues therefore must be totally disaggregated before staining. Usually, enzymatic digestion of the tissue is required to release the component cells. Collagenase and trypsin solutions are commonly used, but the protocol needs to be adjusted to each particular situation. In fact, a protocol that works well with one tissue will not necessarily work for another. Whenever possible, the dead cells must also be eliminated, since their non-specific uptake of fluorochrome-conjugated antibodies could significantly alter results. Once checked, the entire staining procedure can be carried out in test tubes, changing the solutions by centrifugation, and the sample can be directly analysed by the flow cytometer (see protocol 1).

When a microscope is being used, the cells must be placed on a solid support either before staining or just before analysis. If the sample is a single cell suspension (such as blood cells or cultured cell lines growing in suspension), the same protocol used for flow cytometry can be adopted, with the observations being made on a drop of cell suspension placed on a microscope slide. Alternatively, cells can be attached to the slide prior to staining by cytocentrifugation or by means of a pro-adhesive substrate such as poly-L-lysine.

Cytospin has the advantage of flattening the cells, thus making it easier to bring them into focus when taking pictures. For each cell type, the best possible compromise between the physical integrity and yield of adherent cells must be aimed for. If the spin is carried out at too high a speed, the cells will be damaged; if the speed is too low, the cells will not be securely attached to the slide and could be washed off. When the protocol consists of multiple steps, as in the case of multiple indirect immunofluorescence, the chances of losing a portion of the sample at each step are high. In this case, the yield of adherent cells can be increased by pre-coating the slides with a pro-adhesive substrate such as gelatin (see protocol 2). The cytocentrifuge speed may range from 200xg to 800xg. After spinning it is important to check the cell density of the samples under a microscope. If there are either too many or too few cells per field, the results will either be incomprehensible or not representative, and it is advisable to prepare a new set of slides.

If a cytocentrifuge is not available, the cells can be placed on a microscope slide coated with poly-L-lysine. Briefly, 100 µl of a 1 mg/ml poly-L-lysine solution are spread on the surface of a microscope slide by means of a rubber policeman or equivalent device and allowed to dry for 30-60 min at room temperature. A diamond tipped pen is used to circle the area where the cells will be applied; this also helps to delimit the drop of antibody solution. Alternatively, eight-position, multi-test slides are commercially available. Once the slides have dried, 30 µl of a 2x10^5/ml cell suspension in PBS are dropped on each traced circle and the slides are incubated at 4°C for 45 min in a humidified box. The adhesion rate can be checked under the microscope and the optimal incubation time for each cell type can be determined.

Cells growing in adherence can be directly cultured onto sterilised microscope slides. If required, substrates such as fibronectin, collagen, gelatin, or poly-L-lysine can be used to pre-coat glass slides, thus enhancing cell adhesion. Alternatively, adherent cells can be cultured onto special plastic slides provided with culture chambers (e.g. Chamber slides, Nunc). A subconfluent culture produces a good cell density per field while preserving cell integrity. Once the cells have been placed on a solid support, immunostaining is carried out in moist chambers and the slides are washed by immersion in washing buffer.

Fixation and permeabilisation

If the target of immunolocalisation is a cell surface antigen, the cells can be processed without fixation of any kind, using low temperatures (0-4°C) to prevent antibody internalisation. If, due to the cell type, a multiple-step procedure is required that could damage the sample, fixation before staining should be carried out. When the target is an intracellular antigen, cell permeabilisation in addition to cell fixation is necessary to allow free diffusion of antibodies into the cells. In all cases the aim of the procedure must be to preserve both the integrity of the cellular architecture and the antigen/antibody reactivity. The best fixative for a given antigen and assay system must be determined experimentally.

Generally the aldehydes (paraformaldehyde [PFA] and glutaraldehyde [GA]) preserve both the cell structure and its antigenic epitopes. They cross-link the free amino groups in the amino acid side chains of polypeptides. The aldehyde concentration (3-4% PFA; 1% GA; 2% PFA + 0.1% GA) and the incubation time (5-10 min) used to fix cultured cells or cell monolayers usually also partially permeabilise the cell membranes. When permeabilisation is necessary, aldehyde-fixed cells can be washed in PBS and incubated for 5 min with 0.1% Triton X-100 in PBS, or with 0.1% saponin in PBS. Fixation by PFA is partially reversible with prolonged aqueous washes, whereas fixation by GA is not. Therefore, if soluble intracellular antigens are the target, the use of GA + PFA can prevent the loss of proteins during permeabilisation and washing. The only drawback to GA is the presence of free aldehyde groups which could bind antibodies non-specifically. This problem can be overcome by 5 min of washing with a freshly made solution of 0.5 mg/ml sodium borohydride in PBS. Inactivation of the free aldehydes can also be achieved by addition of 0.2 M glycin to the antibody solutions and the washing buffers.

When the antigenic epitope is masked by aldehydes, fixation by organic solvents can efficiently permeabilise the cells while maintaining the antigenic properties of the polypeptides. Organic solvents precipitate cellular proteins and extract the lipid components. The fixation is therefore not reversible, but the cellular architecture may be damaged. The most common protocol uses a 5-min incubation period with absolute methyl alcohol precooled at -20°C. Samples are then kept in PBS until stained. Alternatively, after incubation with methyl alcohol, the slides may be immersed for 5 sec in absolute acetone pre-cooled to -20°C and then air-dried.

These fixation and permeabilisation techniques can also be applied to cells in suspension, provided that care is taken when changing the solutions by centrifugation. The extraction of lipids by organic solvents also makes the cell pellet less visible.

Whichever procedure is used, the samples must be rinsed in PBS before addition of antibodies.

First antibody

In immunofluorescence, as with many other immunochemical techniques, the antibodies can either be labelled directly or they can be detected using a secondary labelled antibody. Generally speaking, a directly labelled primary antibody will produce a cleaner signal and less background, whereas indirect methods generate stronger signals but more background due to the higher number of labelled molecules bound (see below and Figure 1).

Cell staining techniques require that antibodies bind specifically to antigens in the presence of high concentrations of other macromolecules. This means that some antibody preparations, especially those involving polyclonal sera, may show spurious cross-reactions. Furthermore, polyclonal sera will include relatively high concentrations of irrelevant antibodies of unknown

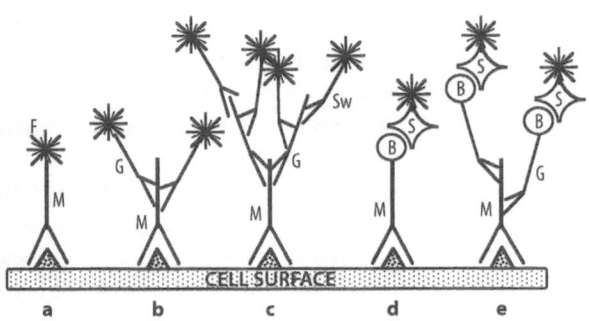

Fig. 1. Basic fluorescence staining principles.
(a) Direct fluorochrome (F) conjugation to a primary mouse (M) antibody;
(b) Indirect staining by a secondary, fluorochrome-labeled goat (G) anti-mouse antibody;
(c) "Sandwich" staining by a tertiary, fluorochrome-labeled swine (Sw) anti-goat antibody;
(d) Indirect staining using biotinylated (B) primary mouse antibody and fluorochrome-labeled streptavidin (S);
(e) Indirect staining using biotinylated secondary goat anti-mouse antibody.

specificity, representing the entire repertoire of the animal at the time of bleeding. These antibodies may create problems by producing intense background when the serum is used at high concentrations.

In many cases, careful titration of the polyclonal serum is sufficient to lower the background. A second possibility is to pre-adsorb the polyclonal serum on cells of the same species, but which do not express the relevant antigen. The best solution, however, is to remove the non-specific antibodies completely by immunoaffinity purification of the antigen-specific antibodies. Carrying out such tedious procedures is justified by the fact that polyclonal sera are specific for a broad range of epitopes on the antigen, including denaturation-resistant epitopes, and therefore also work well on heavily fixed samples.

In contrast, monoclonal antibodies, which usually do not cause as many background problems as polyclonal sera, fail in cell staining if the epitope is hidden or if it is destroyed during fixation. For cell staining purposes, monoclonal antibodies can be used as tissue culture supernatants whereas ascite fluids require careful titration. Like polyclonal sera, ascites can however occasionally generate background problems.

Each set of stained samples should contain at least two controls, i.e. a positive and a negative control. The positive control is used to confirm the effectiveness of the overall staining protocol. The negative control is essential to set the threshold above which cells are to be considered positive for antibody binding. Ideally, the negative control should be an irrelevant antibody of the same immunoglobulin subclass as the test antibody. Suitable control hybridoma cell lines are available from ATCC (Rockville, MD). When such a negative control is not readily available, it may be acceptable to simply use a sample which has not received a primary antibody. Every effort should be made, however, to include a proper negative control since the credibility of the experiment depends upon this. In the case of a polyclonal serum, the ideal negative control would be serum from the same animal collected before the immunisation protocol has been started (pre-immune serum). An unstained sample of cells can also serve as a useful control of the autofluorescence of the cells and the background optical noise. In indirect staining procedures, comparison of the unstained sample with the negative antibody control will also indicate the effectiveness of the secondary antibody blocking.

As noted above, to minimise non-specific fluorescent staining it is essential to take great care in the preparation of the reagents. Aggregates and complexes of immunoglobulins will bind to cells both non-specifically and via Fc receptors. Ultracentrifugation of all reagents immediately prior to staining can reduce this effect. The recommended speed and time may range from 40,000xg for 30 min to 100,000xg for 10 min, but even a few minutes at maximum speed in a microfuge is acceptable.

Fc receptor binding may be excluded by the preparation of (Fab')$_2$ reagents, but this is a tedious and time-consuming procedure that also requires purification of the antibody. To circumvent this problem, Fc receptors may be saturated by pre-treating the sample with 2 mg/ml immunoglobulins or with 10% heat-inactivated serum derived from the same species of cells (in the case of human cells, 10% pooled AB serum will do). Incubation for 20 min at 4°C before adding the first antibody may be sufficient, but we would recommend carrying out the entire staining procedure in the presence of the saturating agent. In the absence of an ideal saturating agent, the following alternatives may be used: 4 mg/ml immunoglobulins from the same species as the second antibody; 2% to 10% fetal calf serum (FCS; heat-inactivated at 56°C for 30 min); or 3% bovine serum albumine (BSA). FCS and BSA, although good competitors for non-specific binding, will not be sufficient to block Fc receptors.

As mentioned above, non-specific staining will in general increase with the concentration of the reagent. All antibody preparations should therefore be titrated against a fixed number of cells in order to find the optimal concentration, i.e. that which provides minimal non-specific binding but effective specific staining. In any case the amount of antibody must not drop below the saturating concentration, which must be determined for each new batch of reagent. Usually 1 to 5 µg/ml antibody, using a working volume of 50 µl/sample and a cell number not exceeding 10^6 cells/sample, will do. The effectiveness of staining is also time-dependent. A 30-min incubation period will usually be adequate, but this may need to be prolonged.

Adjusting the temperature of the reaction can also help to decrease the amount of background. At low temperatures (4°C) only specific binding will be detectable. In addition, a low temperature is mandatory when staining unfixed cells. This blocks the cell metabolism, preventing both internalisation and shedding of bound antibodies and uneven distribution of label (capping). The addition of sodium azide (NaN$_3$ 0.1%) at each step can also block the metabolic activity of the cells.

One last recommendation concerns the washing steps following incubation. At least three such washings for 3-5 min each should be carried out in cold staining buffer containing the saturating agents, plus NaN$_3$ when dealing with unfixed cells. Omission of the washing steps after incubation will result in increased background fluorescence of negative cells and a decrease in the fluorescence of positive cells due to quenching.

When using double labelling, it is possible to design an experiment without too extended a series of steps by adding the first-layer antibodies simultaneously, and subsequently adding the FITC- and PE- (or TRITC-) labelled second antibodies together. In this case the possibility that steric hindrance between the various antibodies may influence binding must first be excluded. Alternatively, each reagent may be added separately and each third washing step may be omitted, except for the final washing after incubation with the fluorochrome-labelled antibodies.

Labelled antibody

One advantage of using indirect staining is that two or more secondary antibody molecules bind to each primary antibody, resulting in a twofold or greater increase in fluorescence intensity compared with directly conjugated antibodies. This is an advantage, particularly if the antigen density is low and binding is difficult to resolve from background.

One disadvantage of indirect staining, however, is that the second reagents are polyclonal antisera, which may exhibit all of the characteristics described in the previous paragraph. Fortunately, carefully adsorbed (Fab')$_2$ fluorochrome-labelled reagents are now commercially available.

It is also possible to use polyclonal antisera produced in different species to build up multiple antibody layers on the primary antibody (Figure 1b, Figure 1c). Each additional layer will increase fluorescence, but will also tend to increase background staining. When planning a multi-layer "sandwich", it is therefore essential to ensure the careful blocking of cross-reactions.

To determine the working dilution of the antibody, in addition to following the manufacturer's indications, the secondary antibody should be titrated against a fixed number of cells incubated with a known saturating concentration of primary antibody.

An alternative labeling and detection method is the biotin/ streptavidin system. The antibody is biotinylated and its binding is detected by the use of streptavidin conjugated to a fluorochrome (Figure 1d, Figure 1e). Streptavidin is highly specific and has a high binding affinity for biotin. Furthermore, one biotin molecule has a valence of four for streptavidin binding. Therefore it enhances even further the signal obtained from the primary antibody and can be used in cases where background staining from the secondary antibody layers is difficult to block. The conjugated streptavidin is titrated and used in exactly the same way as a polyclonal second antibody.

Indirect techniques can also be applied to dual immunofluorescence, but extreme care must be taken to prevent the second antibody from binding to both primary antibodies. When the two primary antibodies are monoclonal antibodies belonging to different immunoglobulin classes or subclasses, it is sufficient to use as the second a fluorochrome-labelled anti-isotype antibody. Alternatively, an unconjugated mouse monoclonal antibody can, for example, be detected with a fluorochrome-labelled anti-mouse antibody. Free anti-mouse antibody is then blocked by the addition of excess mouse immunoglobulins (for instance, 10% heat-inactivated normal mouse serum) before incubation of cells with a second monoclonal antibody conjugated either with the second fluorochrome or with biotin.

All of the suggestions for lowering background provided in the section on primary antibodies apply to the secondary antibodies as well.

Observation of the specimen

The major problem with using fluorochromes as labels is their susceptibility to fading. Although some fluorochromes, such as TRITC, are less prone to do so, quenching of FITC remains a serious problem. This can be minimised by limiting the exposure of the specimen to exciting radiation and by including specific anti-fade reagents, which are commercially available, in the mounting medium. With this problem in mind, stained cells can be fixed and analysed at a later time. Cell pellets can be re-suspended in 2% PFA solution and stored at 4°C in the dark. Microscope slides can be mounted using an aqueous solution such as Gelvatol or Mowiol, or a solution of 1% p-phenylenedia-mine in glycerol/PBS (9 to 1 v/v). In the latter case the samples must be analysed within a few days due to the rapid oxidation rate of p-phenylenediamine.

Once the specimens have been examined, the data can be per-manently stored - in the computer when samples are analysed by a flow cytometer, and in photographic form when samples are analysed by optical microscopy.

The following double-labeling protocols represent two possi-ble solutions to the problem of detection of two different anti-gens in the same specimen. They are paradigmatic and suitable for most antigens and systems.

Future developments

Future developments in immunofluorescence techniques will es-sentially depend on the development of new reagents and tech-nologies for the observation of specimens. Researchers are cur-rently concentrating on the realisation of multi-colour and three-dimensional sample analyses.

Technically, the cardinal problem with multi-colour analysis lies in finding a sufficient number of different probes which are excitable at the same wavelength, but which are able to emit in distinct spectrum areas and therefore remain distinguishable from one another. Recently, considerable attention has been fo-cused on the development of the so-called fluorochrome tandem which consists of the non-covalent association of two fluorescent molecules with complementary spectral characteristics. In such a tandem the first molecule must have an emission range that coincides with the second molecule's excitation range. The ex-citation of the first molecule is therefore transferred directly and without the emission of light to the second molecule, which then emits within its own range. Only two fluorochrome tandems are commercially available at present, Phycoerythrin-Texas Red (PE-TR; excitable at 488 nm and emitting at 610 nm) and Phycoerythrin-Cyanin 5 (PE-Cy5; excitable at 488 nm and emitting at 670 nm). These combinations allow up to four-colour analysis with a single argon laser cytometer, using simultaneous marking with FITC, PE and the two tandems. Other fluorochrome tandems have been synthesised and may become available in the near future, theoretically allowing seven-colour analysis with a dual laser system.

The three-dimensional analysis of a specimen with accurate resolution of the subcellular compartments can be achieved by confocal microscopy. Confocal microscopes use a laser for excitation, computer-controlled motorised nose pieces for focusing, and a digital image analyser to process the plane that is in focus. This ability to exclude all out-of-focus data makes it possible to generate extremely sharp images. Furthermore, the computerised imaging system allows comparison of images at different depths and therefore three-dimensional analysis of the sample. The spread of this particular technology will be limited only by the complex and expensive instrumentation required, since the protocols for immunofluorescence staining are equally applicable to samples being prepared for confocal microscopy. As was the case with flow cytometry, only technical developments resulting in lower costs and simpler procedures will permit widespread application in the future, however.

Troubleshooting

1. Excessive background may be due to any one of the following:
   - too high an antibody concentration.
   - too long an incubation time
   - the absence of competing proteins (i.e. FCS, BSA or Ig) in the staining buffer
   - insufficient washing, especially after fixation and incubation with the fluorochrome-labelled antibody
   - excessive number of dead cells
   - reactivity of the fluorochrome-labeled second antibody with antigenic determinants expressed by the sample (remedy: adsorb the second antibody on the sample)
   - reactivity of some of the Ig of a polyclonal serum with other antigenic determinants expressed by the sample (remedy: affinity purification of the polyclonal primary antibody).

2. Lack of a detectable signal may be due to any one of the following:
   - too low an antibody concentration
   - too short an incubation time
   - in the case of fixed cells, the antigenic determinant may be sensitive to the fixative used
   - if the positive control is satisfactory, then the sample is a true negative.

Subprotocol 1
Dual indirect immunofluorescence on thymocytes:
CD3 antigen versus Bcl2 antigen

▨▨ Materials

Equipment

- Test tubes in conformity with the the requirements of the flow cytometer
- Pipetmen and tips
- Petri dishes
- Fine forceps or syringe needles
- Stainless steel mesh tea-strainer
- Centrifuge
- Flow cytometer

Reagents

- Salts for buffers
- Ethyl alcohol
- Ficoll-Hypaque (Pharmacia)
- Culture medium (e.g. RPMI 1640, Sigma)
- Fetal calf serum (FCS)
- Anti-CD3 antibody (e.g. murine IgG_{2a} anti-CD3 from Pharmingen)
- PE-labelled antibodies (e.g. goat anti-mouse IgG_{2a} from Southern BiotechnologyAssociates)

- Anti-Bcl-2 antibody (e.g. murine IgG_1 anti-Bcl-2 from Dako)
- FITC-labelled antibody (e.g. goat anti-mouse IgG_1 from Southern Biotechnology Associates)

Solutions

- **PBS buffer:**
  0.01 M phosphate (0.25 g $NaH_2PO_4 \cdot H_2O$ + 1.19 g Na_2HPO_4/l) and 0.15 M NaCl, pH 7.2-7.4
- **Staining buffer:** PBS containing 2% FCS and 0.1% NaN_3
- **70% ethyl alcohol**
- **Stock solution of PFA:**
  4% in PBS, pH 7.2-7.4

Preparation

- **Stock solution of 4% PFA in PBS (poison):**
  Dissolve 4 g paraformaldehyde in 100 ml PBS by heating the solution to 50°C while stirring (bear in mind the toxicity of PFA and GA and always use a fume hood) and adding a few drops of 1 M NaOH. Monitor the pH continuously. The solution will remain stable for a few weeks if stored at 4°C in the dark. Just prior to use, dilute in PBS to obtain the required working dilution.
- **Stock solution of 10% NaN_3 (poison):**
  Dissolve 0.1 g NaN_3 in 1 ml water. Keep it at room temperature. Sodium azide is highly poisonous. It blocks the cytochrome electron transport system. Extreme care should be taken in handling.
- **Fetal calf serum (FCS):**
  FCS must be heat-inactivated at 56°C for 30 min before use.

▪ ▪ Procedure

Preparation of the sample (e.g. thymic tissue)

1. Place thymic tissue in a petri dish with a few millilitres of culture medium. Cut into pieces 3-4 mm in size and carefully tease apart each piece with two pairs of fine forceps or with two syringe needles. Pass the cell suspension through

a fine stainless steel mesh tea-strainer. Rescue viable mono-nuclear cells by density gradient centrifugation. Underlay thymocytes with Ficoll-Hypaque (Pharmacia), 10 ml Ficoll every 200x10^6 cells. Centrifuge at 400xg for 20 min at 20°C. Carefully aspirate the opaque layer or mononuclear cells from the interface. Wash twice with culture medium, spinning at 400xg for 10 min at 4°C. Resuspend the pellet to a cell concentration of 10x10^6/ml in staining buffer (PBS, 2% FCS, 0.1% NaN$_3$).

Note: Tissues must be totally disaggregated and dead cells eliminated. The sample should contain as little debris as possible.

2. Load 0.5-1x10^6 cells in a test tube (50-100 µl of the above cell suspension). Add the first antibody (e.g. murine IgG$_{2a}$) detecting the surface antigen (e.g. CD3) at a final concentration of 1-5 µg/ml in staining buffer. The suggested working volume is 100 µl. Incubate for 30 min on ice. **Delivery of the first antibody**

Note: The correct antibody dilution must be determined for each new batch of reagent relative to a fixed number of cells. Never drop below the saturating concentration. Eliminate aggregates by ultracentrifugation. Carry out the entire staining procedure in the presence of a saturating agent. At low temperatures (4°C) only specific binding will be detectable. A 30-min incubation period will usually be adequate, but may need to be prolonged.

3. Add 1-2 ml staining buffer and mix with the cell suspension. Centrifuge at 400xg for 10 min and discard the supernatant. Resuspend the pellet in staining buffer and repeat the washing step twice. **Washing step**

Note: Low temperature is mandatory when staining unfixed cells. The addition of sodium azide during each step will also help to block metabolic activity of the cells.

4. Resuspend the pellet in 100 µl staining buffer containing the fluorescent second antibody (e.g. PE-labelled goat anti-mouse IgG$_{2a}$ heavy chain) at the appropriate dilution. Incubate for 30 min on ice. **Delivery of the PE-labeled second antibody**

Note: The same remarks pertaining to the first antibody apply here. When the two primary antibodies are monoclonal antibodies belonging to different immunoglobulin classes or sub-classes, it will be sufficient to use, as the second antibody, fluor-ochrome-labelled anti-isotype antibodies. Hereafter protect the samples from light by covering with aluminium foil.

Washing step   5. Washing steps as above (step 3).

Fixation   6. Fix the cells by resuspending the pellet in 1% PFA in PBS for 5-10 min on ice (0.5 ml/10^6 cells). Remove PFA by centrifugation at 400xg. Wash once with staining buffer.

Note: Just prior to use, dilute the **PFA stock** in cold PBS to obtain the required working solution. In addition to preserving cell structure during the permeabilisation step, fixation prevents displacement of the first antigen during the staining of the second.

Permeabiliza-   7. Permeabilise by resuspending the pellet in cold 70% ethyl
tion   alcohol (0.5 ml/10^6 cells) for 5 min on ice. Remove the solvent by centrifugation at 400xg. Wash once with staining buffer.

Note: Other permeabilisation techniques can be used (detergents such as Triton X-100, saponin). In the case of anti-Bcl-2 antibody, the use of ethyl alcohol is suggested by the manufacturer.

Delivery of the   8. Incubate cells with an appropriate dilution of the antibody
antibody   (e.g. murine IgG$_1$) detecting the intracellular antigen (e.g.
against the   Bcl-2). Incubate for 30 min on ice.
second anti-
gen   Note: The antibody can be used according to the manufacturer's instructions. The same remarks pertaining to the first antibody against the surface antigen apply here.

Washing step   9. Washing steps as above.

Delivery of   10. Incubate cells with an appropriate dilution of the fluorescent
FITC-labeled   second antibody (e.g. FITC-labeled goat anti-mouse IgG$_1$
second anti-   heavy chain) directed against the second antigen/antibody
body   complex. Incubate for 30 min on ice.

Note: Indirect techniques can also be used in dual immunofluorescence, but great care must be taken to prevent the second antibody from binding to both primary antibodies.

11. Washing steps as above.   Washing step

12. Fix the stained cells with 200 μl of **1% PFA** in PBS. Study the   Fixation and
    sample by flow cytometer within a few days, or place 10 μl on   observation of
    a microscope slide, cover with cover slip and seal with col-   the sample
    ourless nail polish.

Note: If prompt study is not possible, store the samples covered
in aluminium foil at 4°C. Fixation cross-links the antigen/anti-
body complex.

Subprotocol 2
Dual indirect immunofluorescence on T lymphocytes:
CD4 antigen versus IL4 antigen

▨ ▨ Materials

Equipment

- Pipetmen and tips
- Microscope slides and cover slips
- Moist chamber (a box containing dampened paper towels)
- Cytocentrifuge or centrifuge with cytospin adapters
- Scrubbed nylon fibre, 3 deniers, 1.5 inches, type 200
  (Baxter Healthcare, Thetford, Norfolk, UK)
- Optical microscope with UV light and proper filters

Reagents

- Salts for buffers
- Fetal calf serum (FCS, e.g. Bidachem)
- Gelatin (e.g. Sigma)
- Saponin (e.g. Sigma)
- Ficoll-Hypaque (Pharmacia)
- Anti-CD4 antibody (e.g. mouse anti-CD4 from Pharmingen)
- PE-labelled antibodies (e.g. goat anti-mouse from Southern
  Biotechnology Associates)

- Biotinylated anti-IL4 antibody (e.g. anti-IL4 from Pharmingen)
- FITC-labelled streptavidin (e.g. from Southern Biotechnology Associates)

Solutions

- **Saponin 0.1% in PBS**
- **Coating solution:**
  0.25% gelatin, 5 mM $CrK(SO_4)_2$
- **Anti-fade mounting fluid:**
  Either commercial (Vectashield from Vector) or homemade
- **PBS buffer:**
  0.01 M phosphate (0.25 g $NaH_2PO_4 \cdot H_2O$ + 1.19 g Na_2HPO_4/l) and 0.15 M NaCl, pH 7.2-7.4
- **Staining buffer:**
  PBS containing 2% heat-inactivated (56°C, 30 min) FCS and 0.1% NaN_3
- **Stock solution of PFA:**
  4% in PBS, pH 7.2-7.4

Preparation

- **Stock solution of 4% PFA in PBS (poison):**
  Dissolve 4 g paraformaldehyde in 100 ml PBS by heating the solution to 50°C while stirring (bear in mind the toxicity of PFA and GA and always use a fume hood) and adding a few drops of 1 M NaOH. Monitor the pH continuously. The solution will remain stable for a few weeks if stored at 4°C in the dark. Just prior to use, dilute in PBS to obtain the required working dilution.
- **Stock solution of 10% NaN_3 (poison):**
  Dissolve 0.1 g NaN_3 in 1 ml water. Keep it at room temperature. **Sodium azide is highly poisonous.** It blocks the cytochrome electron transport system. Extreme care should be taken in handling.
- **Fetal calf serum (FCS):**
  FCS must be heat-inactivated at 56°C for 30 min before use.

- **Coating solution:**
  Dissolve the gelatin (1.88 g/750 ml water) by heating the solution while stirring (do not boil). When gelatin has been completely dissolved, cool and add 1.88 g chromium alum (chromium potassium sulfate). The solution must be made up fresh for each experiment.
- **Aqueous mounting fluid:**
  Dissolve 100 mg p-phenylenediamine in 10 ml PBS. Add 90 ml glycerol. Store at -20°C protected from light. Use within three months.
- **The nylon fibre column is prepared as follows:**
  Wash the fibre in 0.2 M HCl for 10 min and then thoroughly in water. Pack 0.12 g fibre into a 2-ml syringe. Sterilize by autoclaving. Incubate the column with culture medium containing 10% FCS at 37°C for 10 min and wash it once just before loading with cells.
  Human blood samples should be treated as potential sources of hepatitis viruses and HIV.

Procedure

1. Wash the slides by boiling in soap for 15 min. Rinse thoroughly in distilled water. Dip the slides into freshly made coatsing solution, drain on paper towels and allow to air-dry overnight. Coated slides can be stored in slide boxes.

   Coating of slides

Note: Slides are coated so that the cells will adhere and be retained during the multiple staining steps. Both pre-cleaned microscope slides and glass slides coated with poly-L-lysine are commercially available (e.g. Sigma).

2. Isolate peripheral blood mononuclear cells from heparinised blood (Heparin 10 IU/ml). Dilute the blood with an equal volume of PBS. Underly the blood with Ficoll-Hypaque, 10 ml Ficoll every $3-6 \times 10^8$ total cells (discounting platelets) in ca. 30-mm diameter test tubes. Centrifuge at 400xg for 20 min at 20°C. Carefully rescue the opaque layer at the interface and wash twice with PBS, spinning at 400xg for 10 min.

   Isolation of T-lymphocytes

Note: Lymphocytes can be easily isolated from peripheral blood by sedimentation through a cushion of defined density.

3. Resuspend the cells at 20-40x10^6/ml in culture medium supplemented with 10% FCS and apply to the nylon fibre column. For peripheral blood lymphocytes do not exceed 2.5x10^8 cells/g of fibre. Incubate at 37°C for 30 min. To obtain T cells, elute the nylon fibre with 3-4 ml complete medium at a 1 ml/min flow rate.

Note: T cells can be enriched without challenging their surface molecules using techniques based on the adhesive properties of non T cells: B cells and monocytes bind to nylon fibre whereas the majority of T cells do not.

Cytocentrifugation
4. Mount the microscope slides in a cytocentrifuge with card covers. Add 0.1 ml filtered staining buffer to the reservoir. Spin for 5 min at 400xg to wet the card cover and the slide surface. Load 0.1 ml cell suspension (0.5-1x10^6/ml) in the reservoir and spin again. Check the cell density of the sample under the microscope. Either proceed directly to the fixation step or air-dry the cell monolayer for 15 min.

Note: Centrifuge speed and cell density should be adjusted for each cell type. Avoid having too many or too few cells on the slides. Circle the sample using a diamond-tipped pen; this will also help to delimit the drop of antibody solution.

Fixation
5. Fix the cells by dipping the slides in **1% PFA** in PBS for 10-30 min at 4°C. Alternatively, place one drop of fixing solution on the sample and let sit for 10-30 min. Remove PFA and wash once with staining buffer.

Note: Just prior to use, dilute the PFA stock in cold PBS to obtain the required working solution.

Permeabilization
6. Permeabilise the cells by incubating for 5 min with 0.1% saponin in PBS. Remove the detergent and wash once with staining buffer.

Note: Fixation and permeabilisation occur during the same step when organic solvents are used.

Delivery of the first antibody
7. Add the first antibody (e.g. mouse anti-CD4 antigen) at a final concentration of 1-5 μg/ml in staining buffer. The suggested working volume is 100 μl (1-2 drops). Incubate for 30 min in a moist chamber at 4°C.

Note: The correct antibody dilution needs to be determined for each new batch of reagent relative to a fixed number of cells. Never drop below the saturating concentration. Eliminate aggregates by ultracentrifugation. Carry out the entire staining procedure in the presence of a saturating agent. At low temperatures (4°C) only specific binding will be detectable. A 30-min incubation period will usually be adequate, but may need to be prolonged.

8. Wash twice (for 5 min each time) by dipping the slides in cold staining buffer. Alternatively, gently drop staining buffer onto the slides and then remove using drawn-out Pasteur pipets. After the last washing step, drain the slides on a paper towel.

    Washing step

9. Add 100 μl (1-2 drops) staining buffer containing the fluorescent second antibody (e.g. PE-conjugated anti-mouse serum) at the appropriate dilution and following the manufacturer's instructions. Incubate for 30 min in a moist chamber at 4°C.

    Delivery of the PE-labeled second antibody

Note: The same remarks regarding the first antibody apply here as well. Hereafter, protect the samples from light with aluminium foil.

10. Washing steps as above.

    Washing step

11. Saturate the free second antibody with excess immunoglobulin from the same species as the first antibody. For example, add 100 μl (1-2 drops) of staining buffer containing 10% heat-inactivated normal mouse serum when the first antibody is a murine monoclonal. Incubate for 15 min in a moist chamber at 4°C. Hereafter the additional saturating agent may be added to each incubation.

    Saturation of the second antibody free valence

Note: Indirect techniques can also be used in dual immunofluorescence, but extreme care must be taken to prevent the second antibody from binding to both primary antibodies.

Delivery of
biotin-conju-
gated anti-
body against
the second
antigen

12. Add the biotin-conjugated antibody against the second anti-gen (e.g. biotinylated anti-human IL4) at a final concentra-tion of 1-5 µg/ml in staining buffer. The suggested working volume is 100 µl (1-2 drops). Incubate for 30 min in a moist chamber at 4°C.

Note: The same remarks regarding the first antibody apply here as well.

Washing step

13. Washing steps as above

Delivery of the
FITC-labeled
streptavidin

14. Add 100 µl (1-2 drops) staining buffer containing FITC-la-beled streptavidin at the appropriate dilution, according to the manufacturer's instructions. Incubate for 30 min in a moist chamber at 4°C.

Note: The same remarks concerning the antibodies apply to streptavidin as well.

Washing step

15. Washing steps as above

Fixation (op-
tional) and
observation of
the sample

16. Samples can be fixed in **1% PFA** in PBS, washed and mounted using an aqueous mounting medium such as gly-cerol 90% in PBS containing 0.1% p-phenylenediamine. Add one small drop of glycerol/PBS, cover with a cover slip, re-move excess mounting fluid with blotting paper and seal with colorless nail polish.

Note: If prompt observation is not possible, store the samples protected from light at 4°C. Fixation cross-links the antigen/anti-body complex.

References

1. Harlow E and Lane D. Antibodies. A laboratory manual. Cold Spring Harbor Laboratory 1988.
2. Shapiro HM Practical flow cytometry. 3rd edn. Wiley, New York 1994.
3. Johnstone A and Thorpe R. Immunochemistry in practice. 3rd edn. Blackwell Science, Oxford 1996.
4. Prussin C and Metcalfe D. Detection of intracytoplasmic cytokine using flow cytometry and directly conjugated anticytokine antibodies. J Immunol Methods 1995; 188: 117-128.
5. Roederer M, Kantor AB, Parks DR, Herzenberg LA. Cy7PE and Cy7APC: bright new probes for immunofluorescence. Cytometry 1996; 24: 191-197.

Chapter 6

Immunohistochemistry

GIANCARLO CARBONE

Introduction

The number of antibodies, both polyclonal and monoclonal, of different specificities that have been developed has given a great impulse to immunochemical studies. The availability of these tools has facilitated the localisation of various molecules in tissues and subcellular structures using immunocytochemical techniques. Immunocytochemistry is based on the ability of antibodies to bind a tissue antigen that can subsequently be revealed using direct or indirect techniques.

In direct techniques the preparation is incubated with antibodies directly labeled with either fluorochromes, enzymes, biotin or colloidal gold particles (Fig. 1a). The main advantage of this technique is its rapidity and simplicity of execution. The main drawback is its relatively low sensitivity; indeed, direct techniques can only be used to detect antigens on unfixed cells, i.e. cells in suspension or in sections of frozen tissue. They are not sensitive enough to be used on fixed cytologic or histologic preparations, especially where aldehyde fixatives have been used.

Using indirect techniques, the preparation is first incubated with a primary unlabeled antibody, and subsequently with the secondary antiserum labeled with either fluorochromes, enzymes, biotin or colloidal gold particles (Fig. 1b). Indirect techniques allow the detection of antigens on fixed cytologic preparations and on sections of fixed and paraffin-embedded tissues. They are more sensitive than direct techniques and can detect the antigen even if it has been partially denatured by the fixation procedure.

Giancarlo Carbone, University of Genoa, Department of Experimental Medicine, Histology Unit, Genoa, Italy

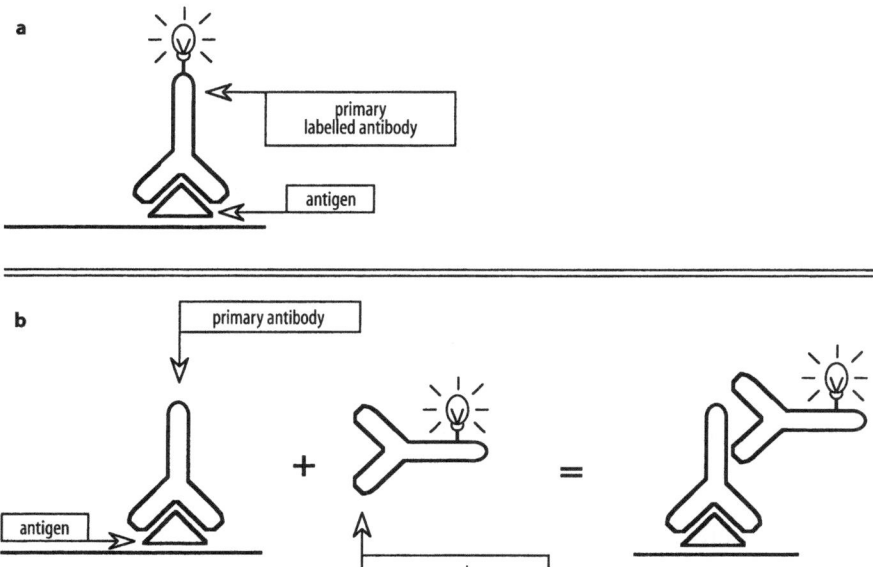

Fig. 1. Detection techniques: a) direct procedure; b) indirect procedure.

The various immunocytochemistry techniques use different labels for the detection of the first antibody: enzymes, fluorochromes or radioisotopes. When enzymes are used (immunoperoxidase or APAAP), a coloured compound is produced and the images are generated by refraction; with fluorochromes, light of various wavelengths is emitted when the appropriate exciting light is used; and with radioactive isotopes, ionising radiation is emitted.

Many different immunocytochemistry techniques are now available, and the advantages and limits of each will be briefly reviewed before the procedures are described. In the protocol section, various possible procedures for each step are outlined. Sometimes different protocols may work equally well in a given situation, and the reader can therefore choose the procedure that seems best suited to the requirements of the experiment.

Sample preparation

The samples can be prepared in many different ways, depending upon the type of sample involved (whole tissues or cells in suspension) and, if the sample is a tissue, its degree of resistance to the reagent preparations. Another factor that must be taken into account is the total length of time required for the procedure.

Cells   Cells can grow either in suspension or adherent to a substrate. Cells in suspension can be placed directly on a slide at the optimal concentration, cytocentrifuged, and then air-dried. The same cytocentrifugation procedure can be used for adherent cells, after first detaching the cells either by scraping or chemically (using EDTA or trypsin-EDTA). This procedure is rapid but does not allow the detection of molecules whose expression is influenced by adhesion to the substrate. For example, adhesion molecules, hormone receptors or cytoskeletal components may change in their level of expression and in their cellular distribution when cells adhere to a substrate. Furthermore, the use of a substrate can produce non-specific staining, especially if polyclonal antibodies are used.

To obtain a preparation of adherent cells suitable for immunocytochemistry studies, the cells are grown on sterile slides coated with collagen. The number of cells to be placed on the slide in a volume of 0.05 ml depends upon the cell growth rate and should be determined separately for each cell line. A high density of cells on the slide can make localisation of the antigen difficult. Overnight incubation in a CO_2 incubator generally helps to optimise cell adhesion and growth. The slides can then be fixed with PLP (paraformaldehyde-lysine-periodate) or methanol/acetone. PLP helps to maintain the integrity of the intracellular structures and the morphology of the cells. Methanol/acetone fixation, in contrast, is simple and rapid, but precipitates cell proteins.

Tissues   Histologic samples can either be placed in a freezing solution (OCT is a commercially available solution often used for this purpose) and frozen at -80°C, or fixed and embedded in paraffin. The freezing procedure is quick and does not alter the antigenic properties of the sample. Tissue sections can then be cut using a pre-cooled microtome.

Fixation and embedding in paraffin, on the contrary, may alter the tertiary structure of the protein molecules and hamper detection of the molecules to be studied, although the morphology of the tissue remains intact. Samples embedded in paraffin can be sectioned using an ordinary microtome at room temperature.

A number of recommendations apply to both procedures and will be reviewed here before the different protocols are described.

1. When handling the bioptic samples, try to avoid any mechanical damage; in particular, do not squeeze them with tweezers.

2. If possible, cut large-sized samples into smaller pieces (0.7 - 1 cm x 0.7 - 1 cm) in order to facilitate the penetration of the fixative.

3. If samples with a very large surface area are needed, macrosections 0.7 - 1 cm in thickness must be cut rather than the usual 15 - 20 mm sections. The fixation procedure must then be carried out in special plastic containers and the reagent consumption will be higher. In order to overcome this problem, in some laboratories large samples are cut into sections following a pre-determined scheme that allows a subsequent reassemblage of the pieces.

4. With organs that have a polarised anatomy such as the kidney or brain, it is advisable to cut cylindric rather than cubic sections in order to be able to easily distinguish the cortical/external side from the medullar/internal side afterwards.The OCT freezing procedure is quite simple and will be described in the protocol section. Below follow some recommendations mainly relating to embedding procedures.

Preparation of the sample for embedding in paraffin:
Fixation, dehydration and clearing

Paraffin embedding must be preceded by fixation, dehydration and clearing of the tissue. These procedures allow the penetration of paraffin into the biopsy, while preserving its morphology.

Fixation   Fixation halts the viable activity of the tissue at a fixed moment in time, just as an image in movement is fixed in a photograph. The fixation step is critical and any error may heavily affect the final result. Two items are of particular importance - the type of fixative and the fixation time used.

Physical fixation is generally accomplished by the application of heat, which can however denature the proteins. Heat fixation is therefore commonly used only with bacteria in microbiology studies. In tissues heat causes irreversible damage to the cells because of their fragility and high water content.

Chemical fixatives are commonly used for histologic studies. They may be applied by perfusion, in vapour form, or by immersion. Fixation by perfusion is accomplished by allowing the liquid fixative to circulate through the vascular bed of the tissue and spread across the capillaries. This method allows the fixation of tissues *in vivo*, since the death of the cells occurs instantaneously with the application of the fixative. Since there is no time lag, the perfusion procedure would be ideal except for the fact that it can only be used in experiments on anaesthesised animals.

Tissues can also be fixed by means of vapours liberated from a heated solution. The fixatives in this case are highly volatile: formaldehyde, glutaraldehyde and osmic acid. This method can be used to fix a tissue sample *in situ* without the necessity of immersing it in a liquid fixative solution, where it could lose its soluble components. However, because of the low rate of penetration of vapors into tissues, this method can only be used with extremely thin sample preparations, such as cell smears.

Fixation by immersion is undoubtedly the most frequently used method for histologic studies. It is performed by dipping pieces of tissues into a liquid fixative for as short a time as possible. The best result is obtained by rapid fixation, which depends on the rate of penetration of the fixative and on the thickness of the tissue sample. Thus, the samples should not be too thick; otherwise some regions may undergo autolytic and/or putrefactive processes before being reached by the fixative. The thickness of the tissue fragments should be proportional to the rate of penetration of the fixative, and as a general rule should not exceed 3 mm.

Usually the speed of penetration will increase with increasing temperature, but it is advisable to carry out the procedure at 4°C, as the degeneration of the tissue is slower at low temperatures while the penetration of the fixative is still quite good. The time required for optimal fixation must be determined separately for each experiment, and will depend on the fixative used, the type of tissue and its thickness, and on the temperature chosen for the fixation step. However, for good fixation the liquid fixative should be removed in most cases within 24 hours unless the fixative chosen also has conservative properties (i.e. formaldehyde). Longer fixation times can produce undesirable effects, such as excessive hardening of the tissue.

Another particularly important point regarding fixation by submersion is the volumetric ratio between the tissue and fixative, which should not be lower than 1:20. It should be kept in mind that some fixatives - e.g. formaldehyde - are consumed during the fixation process and that the release of water from the tissue will further dilute the fixative. The pH and the osmolarity of the fixative should also be taken into account: the pH range should be 7.3 - 7.4 and the osmolarity should be 0.5 oms.

In the protocol section, procedures for the preparation of some fixatives, their characteristics and the incubation times for their use in immunocytochemistry are described. However, it should be kept in mind that incubation times may vary with different tissues and that some trial and error experience will be necessary before the method can be successfully set up and carried out.

Chemical fixation can be carried out using either aldehydes or fixative mixtures.

Aldehyde fixatives cross-link proteins by means of methylene bridges between various amino groups, imino groups, hydroxyl groups, etc. Such bonds could create problems for immunohistochemical reactions, but can be broken by hydrolysis (see below techniques for the unmasking of antigens). Aldehyde fixatives do not affect lipids in any way, but can dissolve carbohydrates since formalin is an aqueous solution. The advantages of aldehyde fixatives are their low cost, ease of handling, and rapid fixation time. They do not cause coarctation, but can harden the tissues, which then become almost elastic. They allow the conservation of lipids, even if they do not fix them. They can also be used to fix large-sized fragments since they have preservative properties. Their main drawback is that they dissolve glycogen and uric acid. Furthermore, their prolonged action can alter the staining properties of the nuclei, masking some protein antigens by the formation of methylene bridges ($CH_2=$). Formaldehyde may form formalin precipitates, although these can be easily removed with abundant washing. Paraformaldehyde also cannot be stored for a long period of time.

Bouin's solution is an excellent fixative for cases in which an overall view of the tissues must be obtained. One of its characteristics is that it preserves the heteropolysaccharides. Although with Bouin's fixative the specimen will shrink considerably in size, it still represents a useful fixative for immunohistochemistry. Muller's solution is particularly suitable for the fixation of nerve and gland tissues. Zenker's solution is an excellent fixative for hematopoietic and connective tissues, and for exocrine gland structures. It maintains the cellular intracytoplasmatic structures intact, but may form precipitates that can mask tissue structures and cause coarctation of the tissue. The precipitates, however, can be washed out by iodised alcohol. Zenker's solution cannot be used on thick sections.

Dhydration and clearing After fixation the tissue must be dehydrated and cleared to allow penetration of the paraffin into the sample. The procedure (described in detail in the protocol section) involves immersing the sample in a series of alcohol solutions of increasing concentration. All of these steps must be performed in Coplin jars using a slide basket.

After dehydration the sample is cleared using substances that are soluble in both absolute alcohol and paraffin. The most commonly used substance is xylene, although it will harden the sample and make it more difficult to cut.

Embedding

This step is carried out using paraffins with different melting points and therefore with different rates of penetration (higher for those with a lower melting point), and a variable consistency and solidity (higher for those with a higher melting point). The paraffins can be placed directly in the oven, taking care never to exceed their point of fusion; otherwise they may become damaged.

In the most widely used procedure (see protocol section), paraffin with a melting point of 50° - 52°C is used for steps I and II, while for step III paraffin with a higher melting point (56° - 58°C) is used. Final embedding is done in paraffin with a higher melting point to which stearic acid 8% w/v has been added in order to provide the section with more elasticity. We use synthetic paraplast-type paraffins. The embedding procedures are described in detail in the protocol section.

Before carrying out immunohistochemical procedures on embedded tissues, a deparaffinisation step followed by a rehydration step must be carried out.

Sectioning

This procedure can be performed in two steps:

1. the coating of the slides with substances that will facilitate the attachment of the sections (optional).

2. the cutting of the samples into thin slices, using a cryostat for frozen samples and a microtome for embedded tissues, followed by the positioning of the sections on the slides.As mentioned above, if the sections are cut from samples embedded in paraffin, a deparaffinisation step followed by a rehydration step is needed before the immunohistochemical procedures can be carried out.

Deparaffinisation and rehydration

In this step the paraffin is removed from the section and the sample is rehydrated, in preparation for the immunohistochemical procedure. All of these steps, as for the dehydration process, must be performed in Coplin jars using a slide basket.

The procedure (described in detail in the protocol section) represents in essence the reverse of the dehydration procedure since it is performed by exposing the sample to a series of alcohol solutions of decreasing concentration.

Unmasking the antigenic sites on samples fixed in formalin and embedded in paraffin by microwave irradiation

Formalin represents the most commonly used fixative, both in diagnostic routine and in the study of the molecules usually expressed by cells. Since formalin causes the formation of cross-links between proteins by means of methylene bridges, some antibodies may not be able to bind to their epitope under these conditions because it has been masked during the fixation procedure.

Exposure to microwaves (generated by an ordinary microwave oven) allows a significant enhancement of the results obtained with many antibodies on routine sections. The efficacy of many reagents when used in conjunction with microwave irradiation has been evaluated. Nevertheless, every laboratory should conduct its own evaluations in order to establish the optimal number of microwave exposures for a given reagent fixation time. The microwave procedure is described in the protocol section.

Immunologic procedures

The samples, either in the form of cells or tissues (the latter either frozen in OCT or embedded in paraffin) can now be incubated with the primary antibody and then with the labeled antibody or immune complex in order to identify their antigenic profile. The general recommendations provided in the other chapters apply; specific recommendations relating to the techniques presented here are described in the protocols.

Subprotocol 1
Immunofluorescence on human mesangial cells

▒ ▒ Materials

Equipment

- Steriliser: e.g. a small autoclave
- Dark glass bottle (or, if not available, a bottle wrapped in aluminium foil)
- Pliers
- Magnetic stirrer
- Multiwell slides (ICN)
- Oven
- Petri dishes
- Laminar flow hood with UV.
- CO_2 incubator
- sterile tubes
- Centrifuge
- Culture chambers
- Cell counter e.g. a haemocytometer slide
- Fluorescence microscope

Reagents

- Lyophylised collagen
- Culture medium
- Salts for buffers: Na_2HPO_4, NaH_2PO_4, $NaCl$, KH_2PO_4
- Trypsin
- EDTA
- Lysin monohydrochloride
- Paraformaldehyde
- Sodium metaperiodate
- Bovine serum albumin (BSA)
- Normal goat serum (NGS)
- Primary antibody: monoclonal anti-smooth muscle actin antibody (e.g. Sigma)
- Labeled antibody: rhodamine conjugate anti-mouse antibody

Solutions

- PBS
- Stock collagen solution
- Trypsin 0.2% w/v -EDTA 0.5% w/v (10x)
- NaOH 2M
- Blocking/diluting solution: BSA 4%, NGS 1% in PBS

Preparation

- **PBS:** Dissolve 7.1 g Na_2HPO_4, 1.36 g KH_2PO_4 and 5.8 g NaCl in 900 ml double-distilled water. Check the pH and, if needed, adjust to 7.4. Adjust the volume to 1 l.
- **Stock collagen solution:** Work under a laminar flow hood and use sterile liquids, tools and recipients.
  Rinse the dark bottle with a small amount of CH_3COOH 0.1 M. Place 15 mg of the lyophilised collagen (using sterilised pliers) in the bottle and add 15 ml CH_3COOH 0.1 M to obtain a 0.1% w/v solution. Stir at room temperature (using a sterile stirring bar) for 3 hours, until the dry collagen is completely dissolved. Add 2.5 ml chloroform and leave overnight at 4°C; do not shake the solution. The following day, two phases will be visible: pipette out and discard the lower phase (containing the chloroform), working under the sterile hood. Store at +4°C, covering the bottle with aluminium foil if you are not using a dark bottle.
  The solution prepared as outlined above is already sterile. It should not be filtered or autoclaved for otherwise it may become damaged.
  At the moment of use, dilute the solution 1:10 in order to obtain a final concentration of 0.01% w/v.
- **Trypsin-EDTA 10x:** Dissolve 20 mg trypsin and 50 mg EDTA in 100 ml PBS. At the moment of use dilute 1:10 in PBS.
- **For PLP fixation** (periodate - lysine - paraformaldeyde):
  - Solution A: Add 1.770 g Na_2HPO_4 to 50 ml double-distilled water.
  - Solution B: Add 0.685 g NaH_2PO_4 to 50 ml double-distilled water.
  - Solution C: Dissolve 1.9 g lysine monohydrochloride in 50 ml solution A, and adjust the pH to 7.4 using solution B. Adjust the volume to 100 ml with double-distilled water.

- Solution D: Dissolve 4 g paraformaldeyde (Merck) in 40 ml double-distilled water at 60°C and clear the solution with a few drops of NaOH 2 M. Let the solution cool to room temperature, filter, aliquote and store at -20°C.
- Just before fixation, mix 7.5 ml solution C, 2.5 ml solution D and 21.4 mg sodium metaperiodate (Aldrich P1314). This solution can be stored for 1 to 2 weeks at 4°C.
- **Blocking/diluting solution:** Add 4 g bovine serum albumin and 1 ml normal goat serum to 90 ml PBS. Dissolve well and then adjust the volume to 100 ml with PBS. Aliquot and store at -20°C. The amount required for blocking and dilution of the antibodies can be thawed as needed.

Procedure

1. Place the multiwell slides in glass petri dishes and sterilise.

2. Dry thoroughly in an oven.

Preparation of the slides: Coating

Note: Work under a laminar flow hood.

3. Cover the slides with collagen solution: 10 µg/cm^2 of solution should be sufficient.

Note: 8-well slides can be covered with 50 µl of collagen 0.01% w/v solution.

4. Cover the petri dishes and store them overnight at 4°C, or at room temperature for several hours.

5. Remove the excess solution from the petri dishes and let them dry overnight in a sterile hood under UV light.

6. The slides can be stored at 4°C, but for no longer than 2 months.

7. At the moment of use, wash with sterile PBS.

8. Aspirate the cell culture medium from the flasks; add 5 ml sterile PBS to each flask and aspirate.

Culture of cells on the coated slides

Note: This method works well with various cell types that adhere as they grow, rather than with cells that grow in suspension.

9. Add 3 ml trypsin-EDTA and incubate for 3-5 min at 37°C. Check under a microscope to see if the cells are detached.

10. Transfer the cell suspension to a sterile tube. Add 3 ml of fresh cell culture medium to the flask, and then transfer to the tube containing the cell suspension. Spin at 300 g for 5 minutes.

Note: A hemocytometer may be used.

11. Aspirate the supernatant; gently resuspend the cell pellet in 1 ml cell culture medium using a sterile pipette. Count the cells.

Note: Avoid dispensing too many cells at once since the localisation of certain antigenic determinants may be difficult when cells are confluent.

12. Dispense the cells onto the pre-coated slides (50 µl per well) and let them grow in culture chambers.

Note: The optimal number of cells will depend on the cell type and its growth rate, approx. 3,000-5,000 per well.

13. Allow the cells to grow in the chamber until they reach a density of 50%.

14. Pipette out the culture medium and wash the slides three times in PBS.

Fixation
15. Prepare the necessary amount of PLP fixative.

16. Fix the cells, covering the slides with the fixative for 20 minutes at 4°C.

Note: This fixation procedure will maintain the integrity of the intracellular structures and the morphology of the cells.

17. Wash twice in PBS.

Note: Each washing should take 5 minutes

18. Submerge the slides in the solution prepared as outlined and leave them for 30 minutes at room temperature.

19. Wash extensively in PBS; at least 3 washings, renewing the PBS each time, will be needed.

20. Cover the slides with a sufficient amount of blocking solution. Incubate for 30 minutes at room temperature.

    Blocking

21. Incubate the slides with the primary antibody (in this case, monoclonal anti-smooth muscle actin antibody) for 2 hours at room temperature or overnight at 4°C. The primary antibody should be diluted 1:400 in blocking solution.

    Incubation with the primary antibody

22. Wash three times in PBS.

Note: The manufacturer's recommendations may serve as general guidelines, but the optimal dilution must be determined for each batch.

23. Incubate with the rhodamin conjugate anti-mouse antibody for 45 minutes at 37°C.

    Incubation with the labeled antibody

24. Wash three times with PBS.

Note: Fluoresceine or rhodamine conjugated antibodies are equally suitable. The manufacturer's recommendations may serve as general guidelines, but the dilution must be optimised for each batch.

25. Mount with glycerol and examine on a fluorescence microscope.

Note: Take photos using a 800 ASA film exposed to 1000 ASA.

Note: For each test two negative controls must be set up and processed along with the other slides: (a) a slide incubated with the diluting solution instead of the primary antibody; (b) a slide incubated with immunoglobulins obtained from a non-immunised animal (same species of the primary antibody) instead of the primary antibody.

Subprotocol 2
Alkaline immunophosphatase (APAAP) on human kidney

▪▪ Materials

Equipment

- Forceps
- Freezer (-80°C)
- Polypropylene tubes
- Plastic supports for OCT
- Tweezers
- Humid box
- pH-meter

Reagents

- Salts: Tris, NaCl, $K_2Cr_2O_7$
- Sulfuric acid
- OCT solution (commercially available)
- Isopentane
- Liquid nitrogen
- Bovine serum albumin (BSA)
- Normal goat serum (NGS)
- Primary antibody: monoclonal anti-cytokeratin antibody (e.g. DAKO)
- Secondary antibody: rabbit anti-mouse antibody
- Monoclonal mouse APAAP (alkaline phosphatase, anti-alkaline phosphatase).
- Naphthol AS-MX phosphate
- Dimethylformamide
- Levamisole
- Fast Red TR
- Weigert's hematoxylin solution (commercially available)
- Glycerol gelatine

Solutions

- TBS
- Sulphochromic solution
- Blocking/diluting solution: BSA 4%, NGS 1% in TBS

Preparation

- **Sulphochromic solution:** Dissolve 100 g potassium dichromate in 900 ml double-distilled water. Add sulfuric acid to a final volume of 1000 ml.

Note: **Warning:** work under fume hood, use a glass beaker and add the acid very slowly as the reaction is highly exothermic. Do not mix the solution with stirring bars and avoid contact with metallic objects. Always wear gloves when working with this solution.

- **TBS:** Dissolve 1.21 g Tris and 8.76 g NaCl in 800 ml distilled water, adjust to pH 7.3 with HCl 37%, and add distilled water to a final volume of 1000 ml.
- **Tris 0.1 M, pH 8.2:** Dissolve 1.211 g Tris in 90 ml double-distilled water, adjust to pH 8.2 with HCl 37%, and add distilled water to a final volume of 100 ml.
- **Blocking/diluting solution:** Add 4 g bovine serum albumin and 1 ml normal goat serum to 90 ml TBS. Dissolve well and then adjust the volume to 100 ml with TBS. Aliquot and store at -20°C. The amount required for blocking and dilution of antibodies can be thawed as needed.
- **Naphthol AS-MX phosphate solution (solution A):** Dissolve 100 mg naphthol AS-MX phosphate in 10 ml DMF. Store at 4°C for a maximum 3 months. Use only glass containers.

Note: Exercise extreme care as naphthol and DMF are toxic. Wear gloves and avoid contact.

- **Chromogenic solution:** dissolve 10 mg Fast Red TR and 2.5 mg Levamisole in 9.8 ml Tris buffer 0.1 M, pH 8.2. Add 200 μl solution A, vortex well, filter with a 0.45 μm filter unit and use immediately.

Note: Eppendorf tubes containing 10 mg Fast Red TR and 2.5 mg Levamisole can be prepared and stored at 4°C for a maximum 6 months.

▪▪ Procedure

Processing the tissue

1. Cut large-sized samples into smaller pieces (approx. 0.7 to 1 cm x 0.7 to 1 cm) in order to facilitate the penetration of the fixative.

Note: When handling the bioptic samples, always try to avoid any mechanical damage; in particular, do not squeeze the samples with tweezers. The sample must **never** be washed in PBS or in any other physiologic solution.

Note: With organs that have a polarised anatomy such as the kidney or brain, it is advisable to cut cylindric rather than cubic sections in order to be able to easily distinguish the cortical/external side from the medullar/internal side afterwards.

Freezing in OCT

2. Prepare a layer of OCT on the plastic support and pre-cool to -20°C.

Note: OCT is a commercially available freezing solution.

3. Pour 25 ml isopentane into a 50-ml polypropylene tube.

4. Dip the tube into liquid nitrogen and allow the isopentane to solidify.

5. Remove the tube with the solidified isopentane and leave it at room temperature until 1/3 of the volume has returned to the liquid phase.

6. Place the sample in the isopentane and leave it there until the isopentane has completely liquified.

7. Using tweezers, gently withdraw the sample and place it on the plastic support, prepared as described above.

8. Cover the sample with fresh OCT and place it in the freezer at -80°C.

9. After one hour check the preparation; if necessary, add more OCT to completely cover the sample.

Note: If one is not planning to continue with the protocol immediately, the sample can be stored at -80°C for several months.

10. Be sure to use clean slides. To clean them properly, submerge the slides in pure sulphochromic solution for 15 minutes.   **Sectioning**

11. Wash the slides using a cleansing solution (make sure that it is RNase-free), and rinse in deionised water 3-4 times. Let the slides air-dry.

Note: Steps 10 and 11 must be performed under a chemical hood since the sulfochromic solution is highly toxic.

12. Cut slices approx. 8-10 μm thick using a cryostat. When cutting each slice, place a clean slide under it; the slice will promptly adhere to the slide.

Note: Before cutting the slices, re-equilibrate the tissue blocks to the same temperature as the cryostat (-30°C).

Note: If one is not planning to proceed immediately with the immunocytochemistry study, the slides can be stored at -20°C in a sealed box. Otherwise, go on to the next step.

13. Let the slides air-dry, protected from dust, for at least 3 hours.

Note: When necessary, after this step one may fix the tissue in methanol/acetone 1:1 at -20°C or in 4% paraformaldehyde in TBS.

14. Wash three times in TBS. Drain and wipe well around the slices.

15. Cover the slides with blocking solution. Incubate for 30 minutes at room temperature   **Blocking**

Note: From this point onward, all incubations must be carried out in a humid box.

16. Incubate with the primary antibody, in this case mouse anti-cytokeratine antibody diluted 1:200 in blocking solution, for 1 hour at 37°C or overnight at 4°C, (Fig. 2A).   **Incubation with the primary antibody**

Note: For all commercially available antibodies, the manufacturer's recommendations may serve as general guidelines for

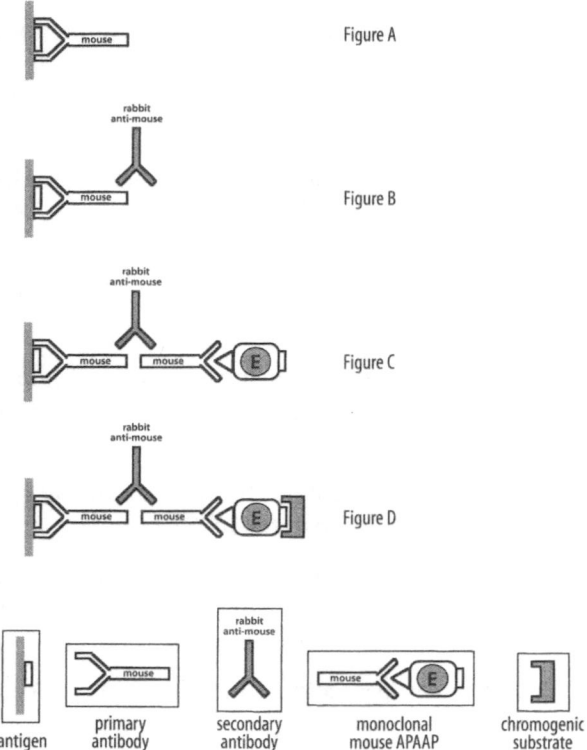

Fig. 2.

the proper dilution, but the optimal dilution must be determined for each batch.

17. Wash three times in TBS.

Incubation with the secondary antibody

18. Incubate with a 1:50 solution of rabbit anti-mouse antibody for 30 minutes at room temperature, (Fig. 2B).

19. Wash three times in TBS.

Incubation with the APAAP complex

20. Incubate with a 1:50 solution of monoclonal mouse APAAP for 30 minutes at room temperature, (Fig. 2C).

21. Wash three times in TBS.

Staining

22. Cover the slides with the chromogenic solution and incubate for 15-30 minutes, (Fig. 2D).

Note: Exercise extreme care as some Fast Red solution components are toxic. Wear gloves and avoid contact.

Note: When the Fast Red substrate interacts with the enzyme, a coloured compound is formed that precipitates in situ.

23. Wash three times in TBS.

24. Differentiate the nuclei by staining with Weigert's haematoxylin for a few seconds.

25. Wash once in TBS for 2 minutes.

26. Wash briefly (for a few seconds) in double-distilled water.

27. Mount in glycerol gelatine for study. Take photos using a microscope with transmitted light.

Since the frozen sample is not washed before being placed in OCT, peroxidase reactions may give false positive results due to endogenous peroxidase present in the blood vessels (see the immunoperoxidase protocols). In contrast, endogenous alkaline phosphatase activity is blocked by the levamisole in the substrate solution.

For each test two negative controls must be set up and processed along with the other slides: (a) a slide incubated with the diluting solution instead of the primary antibody; (b) a slide incubated with immunoglobulins obtained from a non-immunised animal (same species of the primary antibody) instead of the primary antibody.

Subprotocol 3
Immunoperoxidase on human prostate

▓▓ Materials

Equipment

- Forceps
- Tweezers
- Glass container

- Magnetic stirrer
- pH-meter
- Basket for slides
- Coplin jars
- Humid box
- Embedding tray (commercially available)
- Oven
- Microtome
- Thermostated container

Reagents

- Salts: Na_2HPO_4, NaH_2PO_4, $NaCl$, KH_2PO_4, $K_2Cr_2O_7$
- Sulphuric acid
- Paraformaldehyde
- Absolute ethanol
- Methanol
- H_2O_2, 33%
- Xylene
- Paraplast embedding media (paraffin):
  - Paraffin type 1 with melting point ranging from 50° to 52°C
  - Paraffin type 2 with melting point ranging from 56° to 58°C.
- Stearic acid
- APTS (3-aminopropyltriethoxy-silane, e.g. Sigma)
- NaOH
- Bovine serum albumin (BSA)
- Normal goat serum (NGS)
- Primary antibody: mouse anti-collagen IV antibody
- Biotinylated anti-mouse antibody
- Avidin-peroxidase complex
- Diaminobenzidine (DAB)
- $CuSO_4$

Solutions

- PBS
- Sulphochromic solution
- Tris 0.05 M, pH 7.4
- Paraformaldehyde solution

- Alcohol solutions
- DAB solution
- CuSO$_4$ 5%
- H$_2$O$_2$/methanol solution
- Blocking/diluting solution

Preparation

- **PBS:** Dissolve 7.1 g Na$_2$HPO$_4$, 1.36 g KH$_2$PO$_4$ and 5.8 g NaCl in 900 ml double-distilled water. Check the pH and, if needed, adjust to 7.4. Adjust the volume to 1 l.
- **Tris 0.05 M, pH 7.4:** Dissolve 6.055 g Tris in 900 ml water. Adjust the pH to 7.4 with HCl and bring to a final volume of 1 l.
- **Sulphochromic solution:** Dissolve 100 g potassium dichromate in 900 ml double-distilled water. Add sulphuric acid to a final volume of 1000 ml.

Note: Warning: work under fume hood, use a glass beaker and add the acid very slowly as the reaction is highly exothermic. Do not mix the solution with stirring bars and avoid contact with metallic objects. Always wear gloves when working with this solution.

- **Paraformaldehyde solution:** Heat 80 ml PBS to 70°C. Add 4 g paraformaldehyde powder and dissolve well using a magnetic stirrer. Clarify the solution with a few drops of NaOH 1 N (about 70 µl). Cool the solution at room temperature. Adjust the pH to 7.5 and bring to a final volume of 100 ml. Store the solution no longer than one week at 4°C, or aliquot and store at -20°C for no longer than 6 months.

Note: once thawed, the solution cannot be frozen again.

- **Preparation of the embedding media:** The various types of paraffin must be placed in the oven, each at the temperature representing its own melting point, for 24 hours (i.e. 50°- 52°C for paraffin type 1, 56°-58°C for paraffin type 2): with this step any gases present in the paraffin will be eliminated. Paraffin type 2 will be used to embed the sample; add stearic acid until a concentration of 8% w/v is reached. This will provide the paraffin with greater elasticity when it is being cut.

- **Ethanol solutions** at various percentages can be prepared and stored at room temperature in the jars in which the samples are to be submerged, provided that they can be tightly sealed to prevent evaporation.
- **APTS solution:** Add 6 ml APTS to 250 ml methanol.
- **H_2O_2/methanol solution:** add 1 volume of 33% H_2O_2 to 26.5 volumes of pure methanol.
- **Blocking/diluting solution:** Add 4 g bovine serum albumin and 1 ml normal goat serum to 90 ml PBS. Dissolve well and then adjust the volume to 100 ml with PBS. Aliquot and store at -20°C. The amount required for blocking and dilution of antibodies can be thawed as needed.
- **DAB solution:** Dissolve 50 mg DAB in 100 ml Tris 0.05 M. Filter the solution. Just before use add 30 µl of 33% H_2O_2 (2-3 drops).

Note: Exercise great care - DAB is potentially oncogenic. All possible protective measures must be taken in order to avoid direct contact.

- **$CuSO_4$ 5%:** Dissolve 5 g $CuSO_4$ in 100 ml double-distilled water.

▪ ▪ Procedure

Processing the tissue

1. Cut large-sized samples into smaller pieces (approx. 0.7 to 1 cm x 0.7 to 1 cm) in order to facilitate the penetration of the fixative.

Note: When handling the bioptic samples, always try to avoid causing any mechanical damage. In particular, do not squeeze the samples with tweezers. The samples can be washed in PBS or in any other physiologic solution.

Note: With organs that have a polarised anatomy such as the kidney or brain, it is advisable to cut cylindric rather than cubic sections in order to be able to easily distinguish the cortical/external side from the medullar/internal side afterwards.

Fixation by paraformalde-hyde

2. Submerge the tissue in the fixative solution prepared as outlined above, and leave at 4°C for a period ranging from 12 hours to 4-5 days, depending on the thickness of the fragment.

Note: **Exercise extreme care** as paraformaldeyde is toxic. Wear gloves and avoid contact.

Note: A long period of incubation in formaldehyde will not greatly alter the tissues, since formaldehyde is a preservative as well as a fixative.

3. Wash the fixed tissue extensively in PBS. At least 3 washings, lasting 1 hour each, will be necessary.

Note: This washing will remove the paraformaldehyde residues and any possible dark brown precipitates, which represent formalinic pigment produced by the degradation of haemoglobin from the red blood cells.

4. Place the fixed sample in a jar containing the lowest concentration of alcohol. Leave for 1 hour, and then transfer the sample to a second jar containing the same concentration of alcohol for a second incubation. Repeat these steps, exposing

Dehydration and clearing

Table 1. Alcohol series 1 for dehydration and clearing

Solution	Passage	Time
Ethanol 50%	I	1 hour
Ethanol 50%	II	1 hour
Ethanol 70%	I	1 hour
Ethanol 70%	II	1 hour
Ethanol 80%	I	1 hour
Ethanol 80%	II	1 hour
Ethanol 90%	I	1 hour
Ethanol 90%	II	1 hour
Ethanol 95%	I	1 hour
Ethanol 95%	II	1 hour
Absolute ethanol	I	1 hour
Absolute ethanol	II	1 hour
Xylene	I	15 minutes
Xylene	II	15 minutes

the sample to increasing concentrations of alcohol following the series described in the table.

Note: This step allows the paraffin to penetrate into the samples. Use a fresh container for each solution. The lower concentration alcohol solutions may be used more than once, but one must remember that the solutions will gradually be diluted by water originating from the sample. The incubation times may be shortened in some cases.

5. After the second incubation in absolute ethanol, immerse the sample in xylene for 15 minutes. Repeat this incubation step with fresh xylene for another 15 minutes. Then drain the slide and examine it to make sure that the sample is diaphanous, i.e. that the fragment has become translucent.

Note: The absolute ethanol and xylene must be fresh.

Embedding 6. Put the sample in the first jar of paraffin type 1 (there must be sufficient paraffin to cover the sample). Leave for at least 1 hour in the oven at the appropriate melting temperature.

Table 2.

Paraffin step I	50°C-52°C for 1 hour
Paraffin step II	50°C- 52°C for 1 hour
Paraffin step III	56°C- 58°C for 1 hour

Note: The storage and incubations of the paraffins must be carried out at melting temperature. Before the samples are added, the paraffins must be incubated in the oven at their respective melting points for 24 hrs: this will eliminate any gases present in the paraffins.

7. Transfer the sample into the jar containing fresh paraffin type 1. Leave for at least 1 hour in the oven at the same temperature.

Note: This second passage serves to renew the paraffin, since the paraffin in the first jar has been contaminated with xylene.

8. Transfer the sample into the jar containing paraffin type 2. Leave for at least 1 hour in the oven at the higher temperature.

9. Transfer the sample to an embedding tray containing a layer of embedding paraffin. Dip the tray into a jar containing fresh embedding paraffin at melting temperature, remove and allow to solidify at room temperature.

Note: Embedding paraffin: paraffin type II with stearic acid brought to 8%.

Note: During the solidification phase, carefully remove the thin layer of paraffin on the upper side of the block. Store the embedded samples at 4°C.

10. To make sure that the slides are absolutely clean, submerge them in pure sulphochromic solution for 15 minutes.   **Sectioning**

11. Wash the slides in a cleansing solution (be sure that it is RNase-free) and then 3 to 4 times in deionised water. Let the slides air-dry.

Note: This step must be carried out under a chemical hood since the sulphochromic solution is highly toxic.

12. Dip pre-cleaned slides for 3-4 minutes into a small jar containing cold methanol.

Note: Note: If pre-cleaned slides are available, one may begin the embedding procedure with the cold methanol step.

13. Expose the slides to the APTS solution prepared as described. Dip them into the solution 4 times for a total of 30-40 seconds.

Note: Coating the slides with APTS or another solution is necessary to facilitate the adhesion of the sample section.

14. Wash the coated slides in deionised water and dry in an oven at 60°C for 1 hour.

15. Prepare the paraffin block containing the sample. Cut away the excess paraffin.

16. Cut 5-μm slices with a microtome.

17. Put a drop of distilled water on a slide and place it on a flat surface heated to 37°C.

18. Put the section on the drop of water.

19. Wait a few seconds until the section has flattened.

20. Draw off the excess water with a Pasteur pipette and let the section air-dry on the pre-heated flat surface.

Note: Do not leave for too long or else air bubbles will form between the slice and the slide.

21. Leave the slides in a thermostated container for at least 20 hours at 37°C.

Deparaffinisa-
tion and re-
hydration

22. Place the section slides in the first jar of xylene. Leave for 2 minutes, and then transfer the slides to the second jar of xylene for a second incubation. Transfer in a jar containing methanol with 1.2% H_2O_2 and leave for 15 minutes. Repeat the incubation of the samples in a series of alcohol solutions of decreasing concentration, following the indications reported in the table.

Table 3. Alcohol series 2 for rehydration

Solution	Passage	Time
Xylene	I	2 minutes
Xylene	II	2 minutes
Xylene	III	2 minutes
Absolute ethanol	I	2 minutes
Absolute ethanol	II	2 minutes
Ethanol 95%	I	2 minutes
Ethanol 95%	II	2 minutes
Ethanol 80%	I	2 minutes
Ethanol 80%	II	2 minutes
Ethanol 70%	I	2 minutes
Ethanol 70%	II	2 minutes
Ethanol 50%	I	2 minutes
Ethanol 50%	II	2 minutes
Distilled water	I	2 minutes
Distilled water	II	2 minutes

The methanol/H$_2$O$_2$ is needed in order to inhibit endogenous peroxydases.

The absolute ethanol and xylene solutions must be fresh. Use different containers for each solution. The lower concentration alcohol solutions may be used more than once, but keep in mind that with repeated use these solutions will be diluted with water originating from the samples.

The H$_2$O$_2$/methanol solution will inhibit any endogenous peroxidase activity, thus reducing aspecific reactions.

If a fixative containing potassium dichromate has been used, incubation of the slides in Lugol solution for a period of 10 minutes following immersion in 70% alcohol must be carried out.

23. Wash with water, and then three times with PBS.

Note: Store the slides in PBS until you are ready to continue.

24. Drain and carefully wipe dry around the slices. Cover the slides with blocking solution. Incubate for 15 minutes at room temperature.   **Blocking**

Note: From this point onward, all incubations must be performed in a humid box.

25. Incubate with the primary antibody, in this case a mouse anti-collagen IV antibody diluted in blocking solution, for 1 hr at 37°C or overnight at 4°C, (Fig. 3A).   **Incubation with the primary antibody**

Note: For all commercially available antibodies, the manufacturer's recommendations may serve as general guidelines for the dilution, but the optimal dilution must be determined for each batch. For the overnight incubation the antibody concentration can be lower than for the one-hour incubation.

26. Wash three times with PBS.

27. Incubate with biotinylated antibody (anti-mouse or anti-rabbit depending on the primary antibody) for 30 minutes at room temperature.   **Incubation with the biotinylated antibody**

28. Wash three times with PBS, (Fig. 3B).

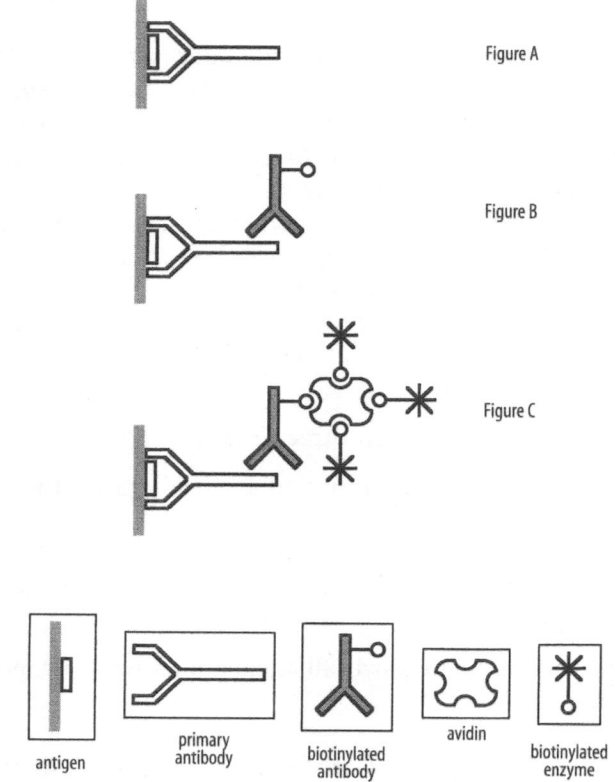

Fig. 3.

antigen	primary antibody	biotinylated antibody	avidin biotinylated enzyme

Incubation with the avidin-peroxydase complex

29. Incubate with avidin - peroxidase complex for 30 minutes at room temperature, (Fig. 3C).

30. Wash three times with PBS

Development

31. Incubate with DAB solution 30 minutes at room temperature.

Note: The DAB solution will react with the enzyme, producing a brown compound that precipitates in situ.

Note: **Exercise extreme care** as DAB is potentially oncogenic. Wear gloves and avoid inhalation. If possible, work under a chemical hood.

32. Wash 3 times with deionised water, 2 minutes for each washing, and then incubate with 5% $CuSO_4$ for 5 minutes at room temperature.

33. Wash with water.

34. Stain the nuclei with haematoxylin for 2-3 seconds.

35. Wash once with PBS for 2 minutes. Then wash for a few seconds in double-distilled water.

36. Mount the slide with glycerol gelatin and observe under a microscope. Take photos under transmitted light.

Note: For each test two negative controls must be set up and processed along with the other slides: (a) a slide incubated with the diluting solution instead of the primary antibody; (b) a slide incubated with immunoglobulins obtained from a non-immunised animal (same species of the primary antibody) instead of the primary antibody.

Subprotocol 4
Preparing slides with cells growing in suspension

Materials

Equipment

- Laminar flow hood with UV
- CO_2 incubator
- Sterile flasks suitable for cell culture
- Sterile tubes
- Cell counter, e.g. hemocytometer slide
- pH-meter
- Centrifuge
- Cytocentrifuge

Reagents

- Salts: Na_2HPO_4, NaCl, KH_2PO_4
- EDTA

Solutions

- PBS
- EDTA 0.02%

Preparation

- **PBS:** Dissolve 7.1 g Na_2HPO_4, 1.36 g KH_2PO_4 and 5.8 g NaCl in 900 ml double-distilled water. Check the pH and, if necessary, adjust to 7.4. Bring the volume to 1 l.
- **EDTA 0.02%:** Dissolve 20 mg EDTA in 100 ml PBS. Check the pH and, if necessary, adjust to 7.4.

▪ ▪ Procedure

Spin down the cells on a microscope slide

1. Collect the cells from the culture flasks and transfer them to a sterile tube.

Note: This method is intended for cells grown in suspension. EDTA facilitates cell dissociation in order to obtain a single cell suspension. The use of trypsin should be avoided since it could damage cellular membranes.

2. Detach the residual cells from the flask walls by adding EDTA 0.02% and incubating at 37°C. Check under a microscope to see if the cells are detached. Collect the liquid and add it to the tube containing the cells.

3. Centrifuge at low speed and then re-suspend the cells in cell culture medium at a concentration of approx. 1,000,000 cells/ml.

4. Mount the slides in the cytocentrifuge. Add 200 µl of cellular suspension to each slide in order to obtain 200,000 cells per spot.

5. Centrifuge at 300 g for 5 minutes.

6. Remove slides from the cytocentrifuge.

7. Let slides air-dry.

8. Proceed with the fixation step.

Subprotocol 5
Preparing slides with cells growing in suspension: fixation

Materials

Equipment

- Glass container

Reagents

- Slides with adherent or spun cells
- Methanol
- Acetone

Solutions

- PBS
- Methanol/acetone 1:1

Preparation

- **PBS:** Dissolve 7.1 g Na_2HPO_4, 1.36 g KH_2PO_4 and 5.8 g NaCl in 900 ml double-distilled water. Check the pH and, if necessary, adjust to 7.4. Adjust the volume to 1 l.
- **Methanol/acetone 1:1:** Mix 50 ml methanol with 50 ml acetone. Cool the solution to -20°C for at least 1 hour before use.

▪▪ Procedure

For PLP fixation see APAAP protocol

Methanol/ace-
tone fixation

1. Submerge the slides in pre-cooled fixative for 2-3 minutes.

2. Let the slides air-dry.

Note: This fixation method is simple to perform but has one significant drawback in that all of the proteins precipitate. Permeabilization is not needed with this method.

3. Wash twice in PBS.

Note: Each washing should take 5 minutes

4. Continue with the immunocytochemistry protocols or otherwise store the slides at 4°C for no longer than 5 days.

Note: Slides with fixed cells may be stored in freezing boxes (e.g. Kartell) at -80°C if necessary.

Subprotocol 6
Coating slides with Poly-L-Lysine

For the APTS coating step see the immunoperoxidase protocol

▪▪ Materials

Equipment

- Containers
- Slides
- Oven

Solutions and reagents

Solutions

- PBS
- Poli-L-lysine (Sigma)
- Sulfuric acid and $K_2Cr_2O_7$ for sulphochromic solution

Preparation

- **PBS:** Dissolve 7.1 g Na_2HPO_4, 1.36 g KH_2PO_4 and 5.8 g NaCl in 900 ml double-distilled water. Check the pH and, if necessary, adjust to 7.4. Bring the volume to 1 l.
- **Sulphochromic solution:** Dissolve 100 g potassium dichromate in 900 ml double-distilled water. Add sulfuric acid to a final volume of 1000 ml.

Note: **Warning: work under fume hood, use a glass beaker and add the acid very slowly as the reaction is highly exothermic.** Do not mix the solution with stirring bars and avoid contact with metallic objects. Always wear gloves when working with this solution.

- **Poly-L-lysine solution:** Add 1 volume of poly-L-Lysine to 9 volumes of PBS

▨▨ Procedure

1. Be sure that the slides are absolutely clean by pouring pure sulphochromic solution into a glass container and submerging the slides in it for 15 minutes. **Cleaning the slides**

2. Wash the slides first in a cleansing solution (be sure that it is RNase-free), and then 3-4 times in deionised water. Let the slides air-dry.

Note: This operation must be carried out under a chemical hood since the sulphochromic solution is highly toxic.

3. Rinse pre-cleaned slides in sterile distilled water for 10 minutes. Drain off the excess water.

Note: If pre-cleaned slides are available, follow the procedure beginning with step 3.

Coating   **4.** Dip the slides in poly-L-lysine solution for 5 minutes

5. Dry in an oven at 60°C for 1 hour.

Note: Store the slides in a dry place, protected from light. Use within six months.

Note: Certain treatments (e.g. microwave irradiation) may cause the sections to detach themselves from the coated slides. In such cases commercially available electrostatic (silanised) slides (e.g. DAKO) may be used.

Subprotocol 7
Fixation procedures

Fixation procedures: paraformaldehyde fixation see immuno-peroxidase protocol

Materials

Fixation by formaldehyde

Equipment

– Glass container
– Magnetic stirrer
– pH-meter

Reagents

– Salts: Na_2HPO_4, NaH_2PO_4, $NaCl$, KH_2PO_4
– Formaldehyde 40% v/v (commercially available)

Solutions

- PBS
- Formaldeyde 4% (formalin)

Preparation

- **PBS:** Dissolve 7.1 g Na_2HPO_4, 1.36 g KH_2PO_4 and 5.8 g NaCl in 900 ml double-distilled water. Check the pH and, if necessary, adjust to 7.4. Bring the volume to 1 l.
- **Formaldehyde 4% (formalin):** Add 10 ml of the commercial solution to 90 ml PBS.

Fixation with Bouin's fixative

Equipment

- Glass containers
- Magnetic stirrer
- pH-meter

Reagents

- Formaldeyde 40% v/v (commercially available)
- Glacial acetic acid
- Picric acid, saturated aqueous solution (commercially available)

Preparation

- **Bouin fixative:**Add 25 ml of commercial formaldehyde solution (40%) and 5 ml acetic acid to 75 ml of saturated picric acid solution. This fixative must be prepared just before use.

Fixation by Muller's solution

Equipment

- Glass containers
- Magnetic stirrer

Reagents

- Potassium dichromate
- Sodium sulphate

Preparation

- **Muller's solution:**Dissolve 2.5 g potassium dichromate and 1 g sodium sulphate in 100 ml distilled water.

Fixation by Zenker's solution

Equipment

- Glass containers
- Magnetic stirrer
- pH-meter

Reagents

- Salts: Na_2HPO_4, NaH_2PO_4, $NaCl$, KH_2PO_4, mercury chloride, potassium iodide, iodine
- Glacial acetic acid
- 70% ethyl alcohol

Solutions

- PBS
- Muller's solution (see above)

- Lugol solution
- 4% Lugol solution in 70% ethyl alcohol

Preparation

- **PBS:** Dissolve 7.1 g Na_2HPO_4, 1.36 g KH_2PO_4 and 5.8 g NaCl in 900 ml double-distilled water. Check the pH and, if necessary, adjust to 7.4. Bring the volume to 1 l.
- **Lugol solution:** Dissolve 2 g potassium iodide in 300 ml double-distilled water, then add 1 g iodine. Store at room temperature in a dark bottle.
- **Zenker solution:** Add 5 g mercury chloride and 5 ml glacial acetic acid to 100 ml Muller's solution. This fixative must be prepared just before use.
- **4% Lugol solution in 70% ethyl alcohol:** Add 4 ml Lugol solution to 96 ml of 70% ethyl alcohol.

Procedure

1. Submerge the tissue in formalin and incubate at 4°C for 12 hours to 4-5 days, depending on the thickness of the sample fragment. **Fixation by formaldehyde**

Note: Formaldehyde is a colourless, irritant gas with an intense smell, and is soluble in water. It is commercially available in the form of a 40% solution. Its penetration rate is approximately 0.8 mm/hr at 4°C.

2. Wash the fixed tissue extensively in PBS; at least 3 washings, for 1 hour each, will be needed.

Note: These washings will remove formaldehyde residues and the brown precipitates (formalinic pigment) produced by the degradation of haemoglobin from red blood cells.

Note: The sample is now ready for dehydration with alcohol.

1. Submerge the tissue in freshly prepared Bouin's fluid and leave it at 4°C for 2 to 12 hours, depending on the thickness of the sample fragment. **Fixation by Bouin's fixative**

2. To eliminate the yellowish colour that follows fixation, wash in a solution of 50% or 70% alcohol.

Note: To accelerate the fixation process, add a few drops of a saturated solution of lithium carbonate. A more dilute solution can be prepared by using 50 ml picric acid and adding 35 ml distilled water.

Note: The sample is now ready for dehydration with alcohol.

Fixation by Muller's solution

1. Submerge the tissue in Muller's solution and leave it at 4°C for about 2 weeks.

Note: The required fixation time is quite long (at least 2 weeks) and the liquid must be changed several times during this period. The samples will become uniformly hard but not excessively coarctated.

Note: The sample is now ready for dehydration with alcohol.

Fixation by Zenker's solution

1. Submerge the tissue in Zenker's solution and leave it at 4°C.

Note: The fixation time required even for small fragments (6-7 mm thick) may vary greatly, ranging from 1 to 24 hours.

2. After the fixation step, the sample must be rehydrated by washing in distilled water for 5-6 hrs with gentle shaking.

3. The sample should now be washed for 10 minutes in Lugol solution prepared as described.

Note: One may now proceed with the series of alcohol incubations.

4. Before carrying out the histochemical reactions on preparations fixed in Zenker's solution and then rehydrated, wash the samples twice (for 5 minutes each time) in PBS or TBS to eliminate the excess iodine from Lugol solution.

Subprotocol 8
Unmasking the antigens

Apply to samples fixed with formaldehyde and subsequently embedded in paraffin. This procedure may improve the recognition of antigens that have been masked by bridges caused by formalin fixation.

▓ ▓ Materials

Equipment

- Microwave-safe plastic container
- Microwave oven (power 650 - 750 W)

Solutions and reagents

- Citric acid monohydrate
- Sodium hydroxide
- Citrate buffer

Preparation

- **Citrate buffer 10 mM:** Dissolve 2.1 citric acid monohydrate in 900 ml distilled water. Adjust the pH to 6.0 with approximately 13 ml sodium hydroxide 2 M and bring the volume to 1 l with distilled water.

▓ ▓ Procedure

1. Deparaffinate the sectioned tissue following the procedure outlined in the immunoperoxidase protocol.

Unmasking
the antigens

Note: If this technique has to be performed, silanised slides (commercially available, e.g. DAKO) must be used, since conventionally coated slides may not ensure adhesion of the slice.

2. Submerge the slides in a container containing citrate buffer. Place the container in a microwave oven.

Note: Note: the container must be microwave-safe and heat-resistant.

3. Irradiate the sections for 2-4 periods of 5 minutes each at 650-700 W.

Note: The solution must reach boiling temperature, and it is important that the sections remain submerged in the buffer. Replenish the buffer in the container after each irradiation, so that the level is maintained.

4. After the last irradiation treatment, remove the container from the microwave oven and let the sections cool to room temperature for 15-20 minutes (they should still be submerged in buffer).

5. Transfer the sections to the buffer solution used for the immunohistochemical procedure (PBS or TBS) and proceed with the protocol.

References

1. Bullock G. and Petrusz P. Techniques in immunocytochemistry; vol 3. New York. Academic Press p. 25-42, 1985.
2. Cordell JL, Falini B, Erber WN, et al. Immunoenzymatic labeling of monoclonal antibodies using immune complexes af alkaline phosphatase and monoclonal anti-alkaline phosphatase (APAAP complexes). J Histochem Cytochem 1984; 32: 219-229.
3. Erber WN and Mason DY. Immunoalkaline phosphatase labeling in hematologic samples. Am J Clin Pathol 1987; 88: 43-50.
4. Mason DY Cordell JL, Abdulazaz Z, Naiem M, Bordenave G. Preparation of peroxidase anti-peroxidase (PAP) complexes for immunohistological labeling of monoclonal antibodies. J Histochem Cytochem 1982; 30:1114-1122.
5. McLean IW, and Nakane PK. Periodate-lysine-paraformaldehyde fixative. A new fixation for immunoelectron microscopy. J Histochem Cytochem 1974; 22: 1077-1083.
6. Willingham MC and Pastan IH. An atlas of immunofluorescence in cultured cells. Academic Press New York p. 1-13, 1995.

Chapter 7

Antibody Specificity analyzed by Peptides synthesized on Cellulose Membranes

PAOLO ROVERO AND DANIELA LUCCHESI

▓ Introduction

It is well known that synthetic peptides represent a powerful tool for the epitope mapping of anti-protein antibodies. Several immunochemical techniques have been developed for this purpose, including immunoblotting and ELISA (see Chapters 2 and 3), but one of the major limitations to the broader application of these techniques is the availability of synthetic peptides. In fact, while more and more laboratories now have their own automated peptide synthesiser, the lack of a large number of synthetic peptides for epitope mapping experiments still constitutes a limiting step in many immunological laboratories.

The "SPOT" method provides a solution to this problem because it is a simple, manual technique for the preparation of multiple peptides directly attached to a solid support in a format suitable for direct antibody assays. The membrane-bound peptide array thus generated is probed with antibodies as in the immunoblot, and the membrane can be subsequently regenerated and re-used several times. Using the SPOT protocol, it is possible to obtain a large variety of paper-bound peptide sets for virtually any kind of immunological test, and also for the study of protein-protein interactions in the broader sense. Thus, overlapping fragments derived from a protein sequence may be used to map linear antigenic determinants or binding sites. Stepwise, N- or C-terminally truncated fragments can be used to elucidate the minimal binding sequence, while the synthesis of substitu-

Paolo Rovero, Department of Pharmaceutical Sciences, University of Salerno, Fisciano, Italy
Daniela Lucchesi, CNR, Peptide Synthesis Laboratory, Institute of Mutagenesis and Differentiation, Pisa, Italy

tion analogues (for instance, the systematic replacement of each residue of a given sequence by Ala in the so-called alanine scan) can help to determine the contribution of individual residues to protein binding. Finally, the SPOT method has been widely exploited in the synthesis of peptide libraries for the *a priori* delineation of peptide binding sequences.

Principle of the SPOT method

The principle on which the SPOT technique is based has been described by Frank and coworkers in several recent papers. A droplet of liquid dispensed onto a porous membrane such as cellulose paper is adsorbed and forms a circular spot. Using a solvent of low volatility containing appropriate reagents, the spot can be considered as a "chemical reactor" anchored to the matrix where chemical reactions can take place. A large number of spots can be distributed on a membrane sheet, to each one of which a distinct reactive substance may be delivered by pipette. In practice, a large set of different peptides can be constructed by dispensing onto each individual spot a drop of solution containing the required activated amino acid. Subsequently, all the standard steps of the synthesis process, i.e. washings, deprotections, etc., are carried out by immersing the entire sheet into the appropriate solvent or reagent. The spotting procedure may be easily automated by means of a robot arm and a dispensing device.

Planning the experiment

A list of the peptides to be synthesised should be drawn up. For a classical epitope mapping experiment, the sequence of the protein under investigation is divided into overlapping peptides, typically with a length of 12 residues. The overlap may be chosen on the basis of how precisely the epitope boundaries should be determined, but the total length of the protein must also be taken into account: a greater overlap (e.g. 10 residues, offset = 2) will give rise to a larger set of peptides which, however, will define the epitopes very precisely. In the example shown in Figure 1 the sequence of human granulocyte-macrophage colony stimulating

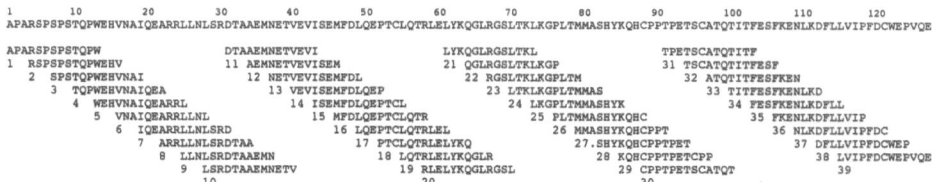

Fig. 1. Sequence of hGM-CSF (from S. Miyatake, T. Otsuka, T. Yokota, F. Lee, K. Arai, EMBO J 1985; 4: 2561) and a possible list of peptides to be prepared for an epitope analysis.

factor (hGM-CSF, 127 residues) has been divided into 39 peptides of 13 residues each, with an offset of 3 (overlap 10).

On the basis of this peptide list, a pipetting protocol can be generated indicating the amino acid to be added to each spot for every cycle (Fig. 1). The spot positions and numbers should be marked in pencil on the membrane.

Membrane preparation

In the first step, a βAla-βAla dipeptide anchor is generated on the cellulose membrane. The presence of this relatively flexible linker will subsequently facilitate both the construction of the peptide chains and, more importantly, the interaction between the peptides and antibodies in the binding assay.

Pure cellulose chromatography paper is chemically derivatised through the uniform esterification of an N^α-Fmoc-protected βAla residue to all of the available hydroxyl functions on the cellulose fibers of the membrane (see Abbreviations). The final substitution level (0.1-0.2 μmol/cm²) will depend on several critical factors, including the thickness of the paper and the purity of the reagents. The efficiency of this step can be monitored spectrophotometrically using bromophenol blue (BPB), an indicator dye that stains free amino groups. The attachment of the first anchor residue is in fact the most critical step of the whole synthetic process. However, functionalised membranes are now commercially available.

In the next step, the Fmoc group is cleaved and the desired array of spots is generated by spotting a second residue of Fmoc-βAla. At this stage, the spotted volume will determine the final size of the spot and thus the scale of the synthesis. Finally, all of the residual amino functions between spots must be capped by acetylation and the Fmoc group on the spots is cleaved.

By this procedure, a βAla-βAla dipeptide anchor is generated at each spot (Fig. 2). It is also possible to introduce a cleavable linker in place of the βAla-βAla anchor, with the aim of cleaving the peptides from the cellulose support. A simple option proposed by Frank is the use of a Boc-Lys-Pro linker, formed using Fmoc-Pro and Boc-Lys(Fmoc), in the two anchoring steps described above. In this case the peptide is released into solution upon treatment with neutral phosphate buffer. This option will not be considered in detail here.

Peptide assembly

Once the linker has been generated, the peptides are synthesised stepwise using a well-established, conventional peptide synthesis protocol based on the "Fmoc/tBu" approach. Briefly, each amino acid residue is introduced ("coupled") onto the growing peptide chain with its N^α-amino group protected by the base-labile Fmoc group, while the trifunctional residues have their side chains protected by acid-labile tBu groups. The coupling is achieved using hydroxybenzotriazole (HOBt) esters of Fmoc amino acid, prepared 30 minutes before use by adding 0.9 M N,N'-diisopropylcarbodiimide (DIC) to a 0.6 M solution of the amino acid in NMP containing 0.9 M HOBt.

Other activation strategies are also possible - for instance, preformed pentafluorophenyl esters which, although easier to handle because they do not require pre-activation, have been found to be less reactive. BPB staining of the free amino functions on the spots enables one to visually monitor each step of the synthesis, since the coupling of an Fmoc-protected amino acid residue onto the growing peptide chains produces a change in the colour of the spot from blue to yellow, while after the subsequent Fmoc deprotection step prior to the next coupling the spot is stained with BPB once again.

The elongation of the peptide chains is achieved by spotting aliquots of the appropriate activated Fmoc-amino acid. If the colour change indicating complete coupling is not obtained within 15 minutes, a second and even a third aliquot can be added until the coupling is driven to completion. However, it should be kept in mind that, although BPB staining provides a very helpful visual aid for monitoring the efficiency of each step of the synthesis process, it does not represent a quantitative technique.

After each coupling, the unreacted amino functions are capped by acetylation, and the Fmoc group is cleaved by treatment with piperidine. The spots are then stained again with BPB before starting the next cycle. At the end of the synthesis, the side chain protecting groups are cleaved by treatment with trifluoroacetic acid (TFA) in the presence of appropriate scavengers, thus leaving one with the peptides in their final form.

Binding assay

The binding of an antibody to paper-bound peptides is most frequently determined by enzyme-labeled antibodies, although other labelling techniques based on radioactive tracers or fluorescent dyes are also available. In the first case, standard enzyme-conjugate/chromogen combinations are used which form water-insoluble coloured products at the binding site. It has been found, however, that with this technique membrane regeneration can be difficult; removal of the insoluble coloured product is time-consuming (see Membrane regeneration) and sometimes complete destaining cannot be achieved.

A much more convenient detection method is based on chemiluminescence. Generally speaking, this procedure is similar to the one used for the dot blot. Membranes are blocked in 5% defatted milk in TTBS. The sera or antibodies are diluted in 2% milk in TBS - 0.1% Tween. After washing, a second antibody labeled with alkaline phosphatase or peroxidase is added. If the antibody is labeled with alkaline phosphatase, the development can be carried out as described in the immunoblotting protocol. If the antibody is labeled with peroxidase, the development is carried out as described in the Protocol section.

Membrane regeneration (stripping)

One of the major advantages of the SPOT technique is that one may re-use the membrane with its bound peptide set many times. In fact, the various reagents involved in the binding assay can be easily removed without damaging the covalently bound peptides.

Possible applications in B cell epitope mapping

In the last four years a number of papers reporting various applications of the SPOT method have been published. As anticipated, the technique was found to be suitable for linear B cell epitope mapping, and monoclonal antibodies raised against several proteins (i.e. interleukin 4, morbillivirus P protein, cytomegalovirus 36/40 K protein, etc.) have been successfully characterised.

In general the SPOT method has proved to be reproducible, and more than 15 different sera can be tested on a single membrane without any apparent loss of spot antigenic activity. When a comparison with the classical ELISA technique was performed, very good overall agreement was observed, although the SPOT method did show a somewhat lower sensitivity. Other reported drawbacks are: (i) the absence of any free charged N- and C- termini, which in some instances can be recognised by antibodies, and (ii) the fact that no analytical control of the peptides is possible.

Interestingly, the SPOT method has been shown to be useful for the epitope analysis of human sera in the clinical setting. Some authors have reported finding no background in tests with sera from autoimmune patients, a problem which has also arisen on occasion using other methodologies.

Further developments: peptide libraries

Yet another powerful application of the SPOT technique has been developed to identify those residues in the sequence of a known epitope that are critical for binding. A cellulose-bound peptide epitope library is constructed synthesising all the individual peptides with each residue of the epitope being replaced in turn by all of the amino acids. The analysis of antibody binding to these peptides provides information on the relative importance of each residue of the epitope.

A further very promising use of the SPOT technology lies in the synthesis of combinatorial peptide libraries on the cellulose membrane. In this application, each spot contains not a single peptide, but a collection of peptides with degenerate amino acid substitutions in one or more positions. In recent years different types of chemically synthesised or biologically generated peptide libraries consisting of millions of distinct molecules have been used for the *a priori* identification of peptides that bind to proteins such as antibodies. The combination of SPOT synthesis with the library technique has made possible the development of cellulose-bound peptide combinatorial libraries that have been successfully used for various assays, including epitope mapping.

Subprotocol 1
Synthesis of overlapping peptides on cellulose membrane to scan human Granulocyte Macrophage-Colony Stimulating Factor (hGM-CSF)

Materials

Equipment

- Micropipette with plastic tips
- Eppendorf tubes
- Timer
- Hair drier
- Rocker table to agitate the solvents or solutions during the washing and incubation steps
- Flat reaction through whose dimensions are slightly larger than those of the membranes used (e.g., petri dish)
- Teflon tube connecting a container to a vacuum line for the collection of solvents and solutions aspirated from the petri dish
- Spectrophotometer
- Whatman chromatography paper (Maidstone, UK)
- Desiccator with P_2O_5
- Molecular sieve (4 Å)

Reagents and solvents

- 1-hydroxybenzotriazole (HOBt)
- 1-methyl-2-pyrrolidinone (NMP)
- Acetic anhydride (Ac_2O; analytical grade)
- Bromophenol blue (BPB)
- Dichloromethane (DCM)
- Diisopropylethylamine (DIEA)
- Dimethylformamide (DMF)
- Ethanol (EtOH; technical grade
- N,N'-diisopropylcarbodiimide (DIC)
- N-methylimidazole (NMI)
- Piperidine (PIP; analytical grade)
- Trifluoroacetic acid (TFA; synthesis grade)
- Triisopropylsilane (TIPS)

Solutions

- **Bromophenol blue (BPB) stock solution:** 10 mg BPB dissolved in 1 mL DMF.
- **Dimethylformamide (DMF):** Must be free of contaminating amines and thus at least of analytical grade. Treat with 4 Å molecular sieves. Amine contamination can be checked by the addition of 10 µL BPB stock solution to 1 mL DMF; the resulting color should be yellow.
- **1-methyl-2-pyrrolidinone (NMP):** Should be of the highest available purity. Treat with molecular sieves until a 1 mL aliquot yields a bright yellow color upon the addition of 5 µL BPB stock solution.
- **N-methylimidazole (NMI):** Distill from solid NaOH and store over molecular sieves at -20° C.
- **Deprotection mixture:** Combine 50% TFA, 3% TIPS, 2% H_2O and 45% DCM.
- **Fmoc-amino acid derivatives:** Side chain protections are Cys(Acm) or Cys(Trt); Asp(OtBu); Glu(OtBu); His(Boc) or His(Trt); Lys(Boc); Asn(Trt); Gln(Trt); Arg(Pmc); Ser(tBu); Thr(tBu); Trp(Boc); Tyr(tBu).

Preparation procedures

– **BPB:** The stock solution should be diluted to 1% with DMF.
– **Fmoc-amino acid derivative stock solutions:** Dissolve each Fmoc amino acid derivative in NMP containing HOBt to yield a 0.6 M solution (1 eq AA + 1.5 eq HOBt). Divide into aliquots and store in labelled Eppendorf tubes at -20°C.

▪ ▪ Procedure

1. Draw up the list of peptides to be prepared (e.g., 39 overlapping 13-mer peptides with an offset of 3 amino acids to scan the hGM-CSF) (see Fig. 1 in the text). Mark the spot positions, 1 cm apart, with a pencil on the cellulose membrane sheet (e.g., Whatman 3MM 6 x 6 cm). The spot numbers refer to the corresponding peptide sequences in the list. Draw up a pipetting sequence specifying the spot number to which each amino acid solution should be delivered (Fig. 3). — *Synthesis planing*

2. Dry the sheet plus two small (1 cm^2) pieces overnight under vacuum on P$_2$O$_5$. Soak the sheet and pieces of membrane in a solution containing 0.6 M Fmoc-βAla, 0.9 M DIC, and 0.9 M NMI in DMF (4 mL/100 cm^2 for 3MM) for 3 hrs. — *Derivatisation of the membrane*

Note: This is a critical step, in which the first anchor component is esterified to the hydroxyl functions of the cellulose.

3. After 3 hours, place one piece of membrane in a small beaker. Wash three times with DMF. Treat with 20% PIP in DMF for 5 min, and wash five times with DMF. Stain repeatedly with 1% (v/v) BPB stock in DMF (1 mL) until the supernatant remains yellow. — *Determination of the substitution level*

Note: Each washing step should take 2 minutes unless otherwise specified.

4. Wash twice with EtOH and dry using a hair drier. Place the piece in a clean small beaker with 5 mL 20% PIP in DMF. Determine the optical density of the deep blue solution and calculate the loading in terms of amino functions per cm^2 (ϵ_{605} = 95.000).

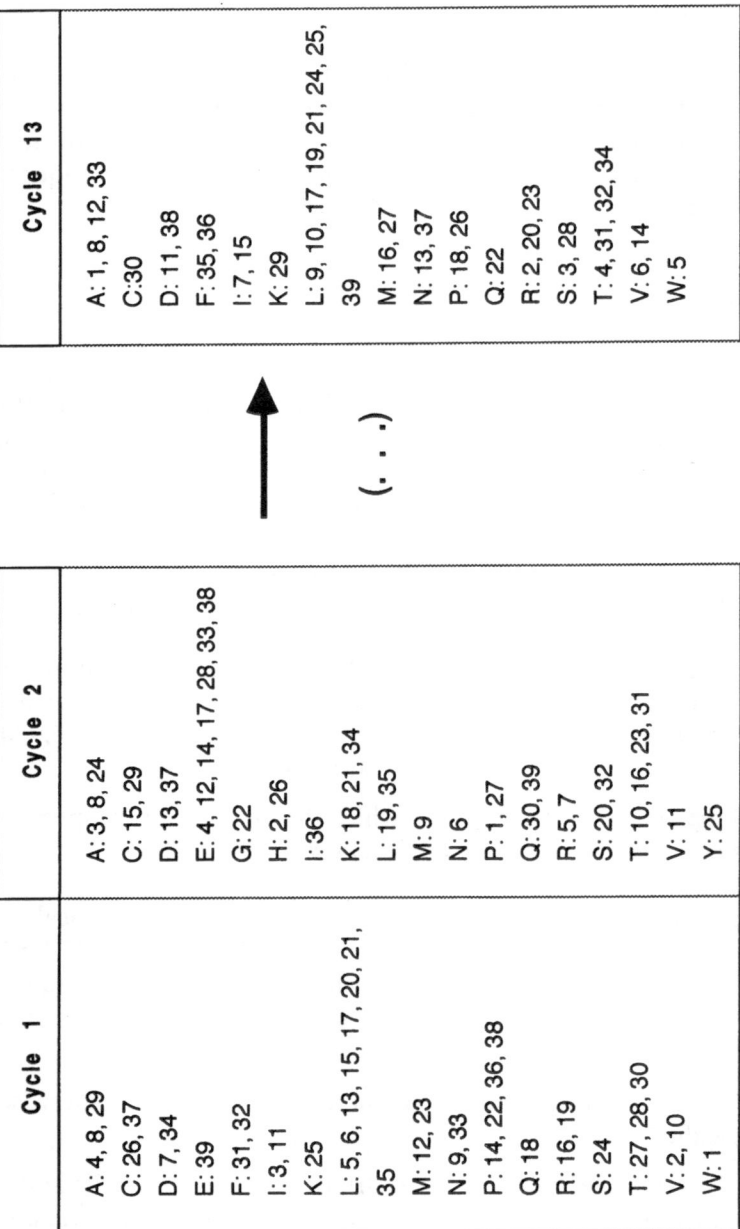

Fig. 2. Pipetting scheme. The amino acids are designated using a single letter code, and are followed by the position numbers of the spots that they should be deposited on for each given cycle.

5. If the loading is < 0.1 μmol/cm^2, wash the membrane three times with DMF and twice with EtOH. Dry.
   Repeat the treatment with the Fmoc-βAla solution described in Steps 2-4. Determine the loading in terms of amino functions per cm^2. If the loading is ≥ 0.1 μmol/cm^2, complete the derivatization of the membrane: wash twice with 2% Ac$_2$O in DMF; acetylate with 2% Ac$_2$O/1% DIEA in DMF for 30 min; wash three times with DMF; treat with 20% PIP in DMF for 20 min; wash five times with DMF; wash three times with EtOH, and finally dry the sheet under vacuum in a desiccator on P$_2$O$_5$.

6. Prepare a solution containing 0.6 M Fmoc-βAla, 0.9 M HOBt, and 0.9 M DIC in NMP; leave for 20 min and then deposit 0.5 μL of this solution on each spot. Leave for 30 min. Repeat this spotting procedure, and wait again for 30 min. Wash twice with 2% (v/v) Ac$_2$O in DMF; acetylate with 2% Ac$_2$O/1% DIEA in DMF for 30 min; wash three times with DMF; treat with 20% PIP in DMF for 5 min; wash five times with DMF; stain with 1% (v/v) BPB stock in DMF; wash twice with EtOH, and then dry the sheet under vacuum in a desiccator on P$_2$O$_5$.

Generation of the spots array

Note: The diameter of the spot obtained by depositing 0.5 μL on 3MM paper will be 4 mm. After BPB staining, the free amino functions on the sheet will appear as distinct blue spots.

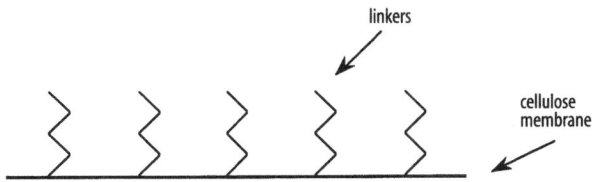

Fig. 3. Arrangement of the βAla linker array on the cellulose membrane.

7. Activate the Fmoc amino acid stock solutions by adding DIC (1.5 eq) and then waiting for 20 min.
   Deposit 1 μL of these solutions on the blue spots, following the pipetting scheme previously drawn up. Leave for 15 min, repeat the spotting and then allow to react for 30 min. Repeat the spotting once again and leave for 15 min. If some of the spots remain blue, repeat the spotting once more.

Assembly of the peptides on the spots

Wash with 5 mL of 2% Ac₂O in DMF for 30 sec, and again with 5 mL of 2% Ac₂O for 2 min. Incubate with 2% Ac₂O/1% DIEA in DMF for 10 minutes until the remaining blue color has completely disappeared. Wash twice with 5 mL DMF. Incubate with 20% PIP in DMF for 5 min. Wash six times with DMF and then stain with 1% (v/v) BPB stock in DMF until the supernatant remains yellow. Wash twice with EtOH. Dry between several layers of filter paper, using a hair dryer on the 'cold' setting. Continue with the elongation cycle (Fig. 4).

Note: The coupling reaction will be followed by a color change in the spots from blue to yellow.

N-terminal acetylation

8. Wash with 2% Ac₂O in DMF for 30 sec, and again with fresh 2% Ac₂O in DMF for 2 min. Incubate for 20 min. Complete acetylation is indicated by the disappearance of the blue color. Wash three times with DMF, twice with EtOH, and then dry using cold air.

Note: Synthetic peptides mimicking B cell determinants represent the fragments of a longer continuous protein chain and therefore should be N-terminally acetylated.

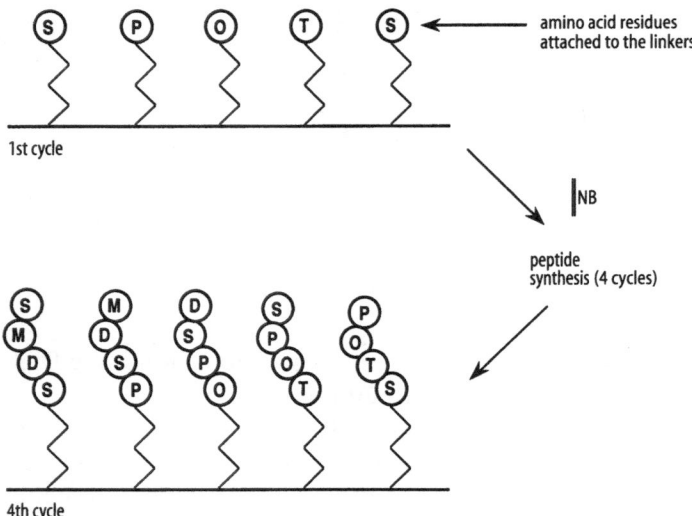

Fig. 4. Assembly of the peptide chains on each linker.

9. Prepare 10 mL of a deprotection solution containing 50% TFA, 3% TIPS, and 2% H_2O in DCM. Incubate the dried membrane under gentle agitation in 5 mL of solution in a petri dish tightly closed to avoid evaporation for 1 hr. Replace the deprotection solution with 5 mL of fresh solution and incubate again for 1 hr under gentle agitation. Wash four times with DCM, three times with DMF and twice with EtOH. Dry the sheet overnight under vacuum in a desiccator on P_2O_5.

Side-chain deprotection

Warning: TFA is highly toxic and this step should be performed under a fume hood.

10. Store at -20° C in a plastic bag.

Storage

Note: The membranes are susceptible to microbe attack if not stored dry in a freezer.

Subprotocol 2
Paper-bound peptide enzyme immunoassay

■ ▨ Materials

Equipment

- Petri dishes
- Timer
- Sonication bath
- Rocker table to agitate the solvents or solutions used in the washing and incubation steps
- Autoradiography film (e.g. Hyperfilm-ECL, Amersham LIFE SCIENCE)
- Film cassette
- Developer/replenisher and fixer/replenisher liquid

Reagents and solutions

- Salts: Trizma base, NaCl, KCl, $NaHCO_3$, $MgCl_2$
- NaOH

- HCl
- EtOH
- Skim milk powder
- NBT
- BCIP
- Urea
- SDS
- DMF
- 2-Mercaptoethanol
- Tween 20
- Acetic acid
- TrisHCl
- TBS
- TTBS
- Blocking buffer
- Carbonate buffer
- Anti-hGM-CSF monoclonal antibody
- Peroxidase-labelled second antibody conjugate: goat anti IgG rabbit/mouse, (e.g., from Boheringer)
- Alkaline phosphatase-labelled second antibody conjugate: goat antimouse IgG(H+L), (e.g., from Biorad)
- Immunodetection reagents (e.g., from Amersham)
- Regeneration buffer
- 70% DMF solution
- NBT stock solution
- BCIP stock solution
- Regeneration Solution A and B

Preparation

- **Tris-buffered saline (TBS):** Add 8.0 g NaCl, 0.2 g KCl and 6.1 g Trizma base in 1 L MilliQ water; adjust pH to 7.0 with concentrated HCl, store at 4°C.
- **TweenTBS (TTBS):** TBS buffer plus 0.1% Tween 20.
- **Blocking buffer:** 5% (w/v) skim milk powder in TTBS.
- **Carbonate buffer:** 0.1 M NaHCO$_3$, 1.0 mM MgCl$_2$. Add 8.4 g NaHCO$_3$, 0.2 g MgCl$_2$ and bring to a final volume of 1 liter with MilliQ water. Adjust the pH to 9.8 with NaOH.
- **Immunodetection reagents (e.g., Amersham):** Just prior to use, mix a volume of detection solution 1 with an equal vo-

lume of detection solution 2 to produce an amount sufficient to cover the membrane.

Note: Direct contact with the reagents should be avoided; the use of gloves is recommended.

- **70% DMF solution:** Mix 0.7 mL DMF with 0.3 mL MilliQ water.
- **NBT stock solution:** Dissolve 50 mg NBT in 1 mL of 70% DMF solution. Store at 4°C.
- **BCIP stock solution:** Dissolve 25 mg BCIP in 1 mL DMF. Store at 4°C.
- **Color development solution:** Just prior to use mix 90 μl NBT stock solution and 90 μL BCIP stock solution in 15 mL of the carbonate buffer, pH 9.8. These substrates will develop an insoluble purple product on the membrane surface following exposure to alkaline phosphatase-coniugated antibodies.
- **Regeneration buffer** (chemiluminescence detection): 50 mM Tris-HCl, 100 mM 2-Mercaptoethanol, 2% (w/v) SDS. Dissolve 10 g SDS, 4.9 g Tris HCl, 3.5 mL 2-Mercaptoethanol in 500 mL MilliQ water and adjust pH to 6.7. Store at 4°C.
- **Regeneration Solution A** (chromogen detection): Urea 8 M, SDS 1%, 2-Mercaptoethanol 0.1 % in MilliQ water. Dissolve 96.1 g urea, 2 g SDS, and 200 μl 2-mercaptoethanol in 200 mL MilliQ water; adjust pH to 7 with acetic acid.
- **Regeneration Solution B** (chromogen detection): 20% acetic acid, 50% EtOH in MilliQ water. Mix 20 mL acetic acid and 50 mL EtOH in 100 mL MilliQ water.

▓ ▒ Procedure

Analysis of epitopes recognised by anti-hGM-CSF monoclonal antibody

1. Shake the membrane in EtOH for 10 min and then wash three times with TBS for 10 min each time.   **Washing step**

Note: Use a large volume of buffer.

2. Incubate the membrane in blocking buffer overnight at 4°C.   **Blocking the membrane**

Note: The efficient blocking of unspecific reactive sites on the membrane and spots is important for the unambiguous detection of specific antibody binding.

Washing step   3. Shake the membrane twice with TTBS for 5 min each time.

Incubation of the first antibody   4. Dilute the first antibody (e.g., anti-hGM-CSF monoclonal antibody) with 2% (w/v) skim milk powder in TTBS and incubate for 3 hrs at room temperature.

Note: The exact dilution of the primary antibody needed to produce optimum results will vary and should be determined for each antibody used.

Washing step   5. Briefly rinse the membrane twice in TTBS, then wash three times with TTBS for 10 min each time.

Incubation of the labelled second antibody   6. Incubate the membrane with the HRP-conjugated second antibody (e.g., goat anti-IgG rabbit/mouse) at the appropriate dilution in 2% (w/v) skim milk powder in TTBS for 90 min at room temperature.

Note: It is important to determine the presence of any non-specific reactions of the peptides with the HRP-labelled secondary antibody. This is done by omitting steps 4 and 5 in a preliminary unspecific binding assay.

Washing step   7. Briefly rinse the membrane twice in TTBS, then wash five times with TTBS for 10 min each time.

Detection step   8. Drain the excess buffer from the membrane. Prepare the detection reagent and add a sufficient amount to cover the peptide side of the membrane, and incubate for 1 min at room temperature without agitation. Drain off the excess detection reagent and place the membrane, peptide side up and sandwiched between two sheets of transparent plastic film, in the film cassette. Switch off the lights and place a sheet of autoradiography film on the membrane, close the cassette and expose the film for 15 sec. Remove the film from the cassette, replace with a sheet of unexposed film and expose for 10 min. Submerge the first sheet of film immediately in developer/replenisher liquid for 2 min, wash the film in water, immerse in fixer/replenisher liquid for 2 min, wash the film in water and air-dry.

Note: Work as quickly as possible to minimize the time lapse between incubating the membrane in the detection reagent and exposing it to the film. The second exposure can vary from 1 min to 1hr, depending on the amount of target on the membrane. If the background is high, the membrane may be re-washed twice for 10 min with TBS and redetected. If over-exposure occurs (due to a high antigen concentration causing high light emission) leave the sheet in the cassette for 5-10 min before re-exposing to film.

9. Wash the membrane twice with TTBS for 10 min each time.   **Washing step**

10. Incubate the membrane with regeneration buffer twice for 15 min each time at 50°C.   **Regeneration of the membrane**

Note: The membrane can be regenerated and re-used several times. If more stringent conditions are required, the incubation can be performed at 70°C.

11. Wash twice with TTBS and twice with EtOH for 10 min each time.   **Washing step**

Note: If the membrane is not to be reprobed, dry between two sheets of paper using a hair dryer set on 'cool', seal in a plastic bag and store at -20°C. If the membrane is to be reprobed, go back to step 1.

Note: The membrane can be probed with AP-labelled antibodies and developed with precipitating substrates. In this case, the following steps should be followed, in place of steps 6–11.

12. Incubate the membrane with the AP-conjugated second antibody (e.g., goat anti-IgG(H+L) mouse) at the appropriate dilution in 2% (w/v) skim milk powder in TTBS for 90 min at room temperature.   **Incubation of the labelled second antibody**

13. Wash three times with TTBS for 10 min each time and then wash once with carbonate buffer for 10 min.   **Washing step**

14. Submerge the membrane in the color development solution for 4-5 min. Immerse the membrane in Milli Q water to stop the reaction. Drain off the water and rinse once again with Milli Q water.   **Detection step**

Note: Keep the membrane wet; if it dries out, the proteins may denature and become difficult to remove.

Washing step
15. Wash the membrane twice with water, three times with DMF and three times with Milli Q water for 10 min each time.

Note: Perform the second DMF washing under sonication until the color of the spot signals has disappeared.

Regeneration of the membrane
16. Incubate the membrane with solution A three times for 10 min each time at 40°C, and then with solution B three times for 10 min each time.

Note: If more stringent conditions are required, the incubation with solution A can be performed at 70°C.

Washing step
17. Wash twice with H_2O and then twice with ethanol for 10 min each time.

Note: If the membrane is not to be reprobed, dry between two sheets of paper using a hair dryer set on 'cool', seal in a plastic bag and store at -20°C. If the membrane is to be reprobed, go back to step 1.

References

M.H.V. van Regenmortel, J.P. Briand, S. Muller, S. Plauèè: Synthetic Poly-peptides as Antigens, Elsevier, Amsterdam (1988). Series: laboratory techniques in biochemistry and molecular biology, general editors: R. H. Burdon and P. H. van Knippenberg.

Frank R. Spot-synthesis: an easy technique for the positionally addressable, parallel chemical synthesis on a membrane support. Tetrahedron 1992; 48: 9217-9232.

Frank R, Overwin H. Epitopes analysis with arrays of synthetic peptides prepared on cellulose membranes. Methods in Molecular Biology 1996; 66: 149-169.

Halimi H, Dumortier H, Briand JP, Muller S. Comparison of two different methods using overlapping synthetic peptides for localizing linear B cell epitopes in the U1 snRNP-C autoantigen. J Immunol Methods 1996; 199: 77-85.

Gao B, Esnouf MP. Multiple interactive residues of recognition. J. Immunol 1996; 157: 183-188.

Schneider-Mergener J, Kramer A, Reineke U. Cellulose-bound peptide libraries as a tool to study molecular recognition. In Cortese R ed, Combinatorial Libraries. Berlin: W. de Gruyter , 1995: 53-68.

Abbreviations

Ac_2O	acetic anhydride
Acm	acetamidomethyl
AP	alkaline phosphatase
BCIP	5-bromo-4-chloro-3-indolyl phosphate p-toluidine salt
Boc	tert-butyloxycarbonyl
BPB	bromophenol blue
DCM	dichloromethane
DIC	N,N'-diisopropylcarbodiimide
DIEA	diisopropylethylamine
DMF	dimethylformamide
ECL	enhanced chemiluminescence
ELISA	enzyme-linked immunosorbent assay
eq	equivalent
EtOH	ethanol
Fmoc	9-fluorenylmethyloxycarbonyl
HCl	hydrogen chloride
hGM-CSF	human granulocyte-macrophage colony stimulating factor
HOBt	1-hydroxybenzotriazole
HRP	horse radish peroxidase
KCl	potassium chloride
$MgCl_2$	magnesium chloride
NaCl	sodium chloride

$NaHCO_3$	sodium hydrogen carbonate
NaOH	sodium hydroxyde
NMI	N-methylimidazole
NMP	1-methyl-2-pyrrolidinone
NBT	p-nitro blue tetrazolium chloride
P_2O_5	phosphorus pentoxide
PIP	piperidine
Pmc	2,2,5,7,8-pentamethylchroman-6-sulfonyl
SDS	sodium dodecyl sulfate
TBS	Tris-buffered saline
tBu	tert-butyl
TFA	trifluoroacetic acid
TIPS	triisopropylsilane
Tris	tris(hydroxymethyl)aminomethane
TTBS	Tween Tris-buffered saline
Trt	triphenylmethyl

Appendix 1

Ammonium sulphate precipitation

LAURA CAPONI

Introduction

One of the most common methods used for the purification of antibodies from serum or from cell culture medium is ammonium sulphate precipitation, which is a very economical and simple procedure. It represents a somewhat "raw" method since the antibodies obtained are generally contaminated with proteins of smaller size present in the sample, but it is sufficient for most purposes and in any case can be used as a preliminary antibody purification step before ion-exchange or affinity purification chromatography.

The underlying principle of ammonium sulphate precipitation is quite simple. Even if soluble in aqueous medium, immunoglobulins (like most other proteins) exhibit hydrophobic areas at their surface that constrain a certain number of water molecules to a forced orientation. The addition of increasing concentrations of salt ions means that increasing numbers of water molecules will be required for solvation, thus forcing the proteins to interact with one another through their hydrophobic regions. In other words, the progressive addition of salt reduces the solubility of the proteins, causing them to aggregate via hydrophobic bonds and then precipitate. The larger the amount of hydrophobic areas exposed by specific proteins, the less salt must be added to reduce their solubility and induce precipitation.

Ammonium sulphate precipitation can be carried out adding the salt directly to the protein solution, but in a small-scale system it may be more convenient to add an appropriate amount of

Laura Caponi, University of Pisa, Department of Internal Medicine, Clinical Immunology Unit, Pisa, Italy

saturated ammonium sulphate solution to reach the final salt concentration required; for antibody precipitation this concentration is 50%.

Procedure

1. Prepare a saturated solution of ammonium sulphate by dissolving 704 g $(NH_4)_2SO_4$ in distilled water and adjusting the volume to 1 liter. This solution can be stored at 4°C.

2. Centrifuge the sample containing the antibodies at 3000 g for 20 minutes in order to precipitate all of the insoluble material.

3. Save and quantify the supernatant and dispense it into a container of suitable size in an ice-bath placed on a magnetic stirrer. Drop a stirring bar into the sample. Slowly add a volume of saturated ammonium sulphate equal to the volume of the supernatant with continuous stirring.

Note: The stirring should not be too energetic or the proteins could be mechanically damaged; foaming must be avoided since it could denature the proteins. The ammonium sulphate solution must be added slowly to allow mixing without the formation of salt nuclei at percentages higher than required.

4. When the ammonium sulphate is perfectly mixed with the protein solution, incubate at 4°C.

Note: Gentle stirring is recommended also in order to facilitate the interaction among the salt ions and proteins and among the proteins themselves. The minimum time required for the reaction is 30 minutes, but if possible 3 hours of incubation is preferable.

5. Centrifuge the solution at 3000 g for 30 minutes at 4°C.

Note: Discard the supernatant and dissolve the precipitate in physiological buffer. Using low amounts of buffer, high antibody concentrations can be obtained; ammonium sulphate precipitation can in fact be considered both a purification and a concentration procedure. The protein precipitate will now contain a considerable amount of ammonium sulphate ions that can be removed by dialysis or gel filtration on a desalting column.

6. Quantification can be performed using one of the many available protein estimation methods (BCA, Lowry, Coomassie, etc.) and suitable calibration standards.

Ammonium sulphate precipitation can be carried out in a single step as described above, but sometimes it is preferable to perform a preliminary precipitation step at a lower salt concentration (e.g., 25%) in order to reduce the contamination (by other proteins) of the final antibody precipitate obtained. The precipitate obtained at the lower concentration can be discarded because it usually contains a low amount of antibodies but a considerable amount of other proteins. The supernatant can then be saved and quantified, and the correct amount of saturated salt solution required to bring the concentration to 50% can be added. With this procedure the purity may be higher but the yield lower, since an additional step resulting in a small loss of antibodies has been introduced.

Appendix 2

SDS-PAGE: Sodium Dodecyl Sulphate Polyacrylamide Gel Electrophoresis

LAURA CAPONI

Materials

- **Acrylamide 30%:** Dissolve 29.2 g acrylamide and 0.8 g bis-acrylamide in 70 ml distilled water; bring to a volume of 100 ml. Filter and de-gas the solution. Store at 4°C in a dark bottle.

Warning: Acrylamide is extremely toxic. Wear gloves and a mask when handling the powder.

- **Ammonium persulphate 10%:** This reagent must be prepared fresh each time. Alternatively it can be prepared, aliquoted, stored at -20°C and thawed when needed.
- **Tris 0.5 M, pH 6.8:** Dissolve 30.275 g in 400 ml distilled water. Bring to pH 6.8 with HCl and then add water to a volume of 500 ml. Store at 4°C.
- **Tris 1.5 M, pH 8.8:** Dissolve 90.82 g in 400 ml distilled water. Bring to pH 6.8 with HCl and then add water to a volume of 500 ml. Store at 4°C.
- **SDS 10%:** Dissolve 1 g sodium dodecyl sulphate in 10 ml water. Store at room temperature.

Warning: SDS is toxic. Wear a mask when handling the powder.

- **Sample buffer 4x:** Dissolve 400 mg SDS in 2.5.ml Tris 0.5M, pH 6.8. Add 2.3 ml glycerol and 1 mg bromophenol blue. Store at room temperature.

Note: When the sample has to be reduced, prepare a small amount of reduced sample buffer (1 volume of β-mercaptoethanol + 4 volumes of sample buffer 4x).

Laura Caponi, University of Pisa, Department of Internal Medicine, Clinical Immunology Unit, Pisa, Italy

- **Preparation of the sample:** add half a volume of reduced or unreduced sample buffer 4x to 1 volume of the protein sample. Boil for 3 minutes and then let cool at room temperature before loading.

Procedure

1. Assemble the apparatus following the manufacturer's instructions. The glass plates must be clean and handled with gloves. Clips may be used (see figure) to hold the glass plates and the two spacers (thickness range: 0.5 - 1.5 mm) firmly together.
   The volumes of the single reagents can easily be modified depending on the volumes of running and stacking gels needed.

2. Following the indications given in the table, prepare the acrylamide mixtures using the stock solutions. Do not add ammonium persulphate and Temed until you are ready to pour the solution between the glass plates since the polymerisation reaction will start immediately after their addition.

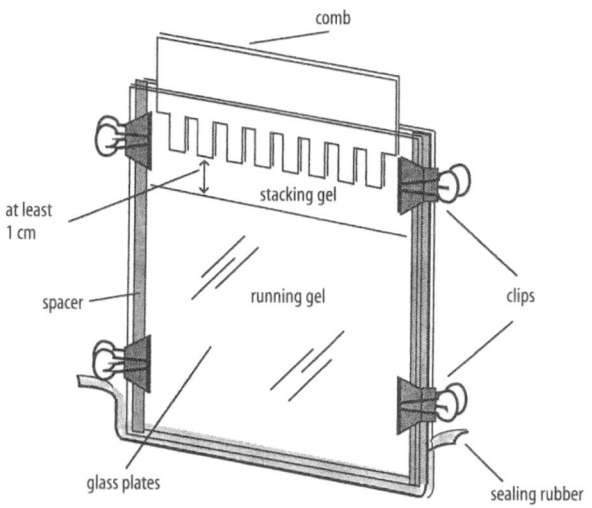

Fig. 1.

Table 1. Running gel

	40 ml 7.5%	40 ml 10%	40 ml 12.5%	40 ml 15%
H_2O ml	19.4	16.2	13	9.6
Tris, pH 8.8 ml	10	10	10	10
Acrylamide 30% ml	10	13.2	16.4	19.8
SDS10% ml	0.4	0.4	0.4	0.4
Ammonium persulphate 10% μl	20	20	20	20
Temed μl	20	20	20	20

Table 2. Stacking gel

	10 ml
H_2O ml	6.1
Tris, pH 6.8 ml	2.5
Acrylamide 30% ml	1.3
SDS 10% ml	0.1
Ammonium persulphate 10% μl	50
Temed μl	10

Note: Make sure that the plate + spacer assemblage does not leak at the base or along the sides. A 1.5% agar solution, which solidifies at room temperature, can be used to seal the edges. Warm the agar and spread it evenly along the sides of the assemblage.

3. Add ammonium persulphate and Temed to the running gel solution, mix gently and pour the solution between the glass plates. Using a Pasteur pipette, immediately gently spread a thin layer of water over the gel solution. Let the gel polymerise at room temperature for at least 45 minutes. When the edge between acrylamide and water becomes visible, the water can be poured off.

4. Add ammonium persulphate and Temed to the stacking gel solution, pour it onto the running gel, and place an appropri-

ately sized comb. The stacking gel will polymerise in a shorter time than the running gel. At least 1 cm of stacking gel must be present between the bottom of the comb tooth and the edge of the running gel. When the stacking gel has polymerised, the comb may be carefully removed and the grooves left by the teeth of the comb can be washed with running buffer solution in order to remove the unpolymerised acrylamide.

5. The apparatus can now be placed in the gel box with the upper and lower reservoirs filled with running buffer. Load the samples and start the run at low voltage (90 - 100 volts). When the dye front reaches the separating gel, the voltage can be increased. The gel can be run either at a constant voltage or at constant current. If the voltage or current is too high, the gel may heat up, with a resulting distortion of the protein bands. If a cooling unit is available, a higher voltage/current may be employed.

When the dye front has reached the bottom of the plates, the power can be turned off and the plates can be removed and disassembled. The separated proteins in the gel can then be transferred to nitrocellulose (as described in the immunoblotting chapter) or directly stained with Coomassie.

Appendix 3

Protein A and Protein G

LAURA CAPONI

Protein A and protein G are proteins constitutive of bacterial cell wall. They are able to bind the IgG Fc fragment, leaving unaltered the binding capacity of antibodies. For this property they represent an invaluable immunochemical tool for isolation of immunoglobulins. Protein A (but not protein G) can bind other antibody classes.

Table 1. Binding ability of IgG subclasses in human and mouse

	S. aureus protein A	Streptococci protein G
Human IgG1	+++	+++
Human IgG2	+++	+++
Human IgG3	-	+++
Human IgG4	+++	+++
Mouse IgG1	0	0
Mouse IgG2a	+++	+++
Mouse IgG2b	+++	+++
Mouse IgG3	+++	+++

References

Purification and some properties of streptococcal protein G, a novel IgG-binding reagent. Björck L and Kronvall G. J Immunol 1984. 133 (2) 969-974.

Protein A of Staphylococcus aureus and related immunoglobulin receptors produced by Streptococci and Pneumonococci. Langone JJ. Advances in immunology. Vol 32, 1982, 157-251.

Laura Caponi, University of Pisa, Department of Internal Medicine, Clinical Immunology Unit, Pisa, Italy

Subject Index